Simplifying Application Development with Kotlin Multiplatform Mobile

Write robust native applications for iOS and Android efficiently

Róbert Nagy

BIRMINGHAM—MUMBAI

Simplifying Application Development with Kotlin Multiplatform Mobile

Associate Group Product Manager: Rohit Rajkumar
Publishing Product Manager: Shalita Aranha
Senior Editor: Mark Dsouza
Content Development Editor: Divya Vijayan
Technical Editor: Joseph Aloocaran
Copy Editor: Safis Editing
Project Coordinator: Rashika Ba
Proofreader: Safis Editing
Indexer: Manju Arasan
Production Designer: Ponraj Dhandapani
Marketing Coordinator: Elizabeth Varghese

First published: March 2022

Production reference: 1280122

Published by Packt Publishing Ltd.
Livery Place
35 Livery Street
Birmingham
B3 2PB, UK.

ISBN 978-1-80181-258-0

www.packt.com

Contributors

About the author

Róbert Nagy is a Senior Android Developer at Octopus Energy. He is an Android and Kotlin developer with a Bachelor of Science in Computer Science. He has designed, developed, and maintained multiple sophisticated Android apps ranging from 100K+ downloads to 10M+ in the financial, IoT, health, social, and energy industries. Some projects that he has been a part of include a social platform for kids, a lightning system controller, and Bloom and Wild.

About the reviewers

John O'Reilly is a Kotlin Google Developer Expert with over 30 years of hands-on software development experience. He has been developing Android apps since 2010, and he worked on server-side Java applications in the 2000s and desktop clients in the 1990s. He has taken a keen interest over the last few years in all things Kotlin Multiplatform and, in particular, when combined with use of declarative UI frameworks such as Jetpack Compose and SwiftUI.

Yev Kanivets is an experienced native mobile developer working on both Android and iOS applications since 2014. He became an early adopter of Kotlin Multiplatform in 2019 sharing everything up to the presentation level for a multitude of production apps.

His work experience includes multiple consultancies and product companies, as well as holding the CTO and co-founder positions at xorum.io, a company dedicated to Kotlin Multiplatform solutions. There he is responsible for tech decisions, tech talks (.droidcon, VKUG, and so on), workshops, articles for the Medium blog, and other tech stuff.

Table of Contents

2

Exploring the Three Compilers of Kotlin Multiplatform

3

Introducing Kotlin for Swift Developers

Section 2 - Code Sharing between Android and iOS

4

Introducing the KMM Learning Project

5
Writing Shared Code

6
Writing the Android Consumer App

7
Writing an iOS Consumer App

Section 3 - Supercharging Yourself for the Next Steps

8
Exploring Tips and Best Practices

9
Integrating KMM into Existing Android and iOS Apps

10
Summary and Your Next Steps

Other Books You May Enjoy

Preface

Kotlin Multiplatform is the new tool in the toolkit of native developers. It provides gradual code-sharing capabilities between multiple platforms, including Android and iOS.

This book introduces this new technology from the perspective of mobile development. It gives you the necessary knowledge to extend your development toolset and build a more effective native development process.

After reading this book, you should have a clear understanding of Kotlin Multiplatform's strengths and how you can leverage the tool to build mobile applications more effectively.

Who this book is for

This book is for native Android and iOS developers who want to build high-quality apps using an efficient development process. Knowledge of the framework and the languages used is necessary, that is, Android with Java or Kotlin and iOS with Objective-C or Swift. For Swift developers, the book assumes no knowledge of Kotlin as this will be covered in the context of Swift.

What this book covers

Chapter 1, The Battle Between Native, Cross-Platform, and Multiplatform, compares the available cross-platform frameworks with native frameworks and introduces Kotlin Multiplatform.

Chapter 2, Exploring the Three Compilers of Kotlin Multiplatform, describes the architecture of Kotlin Multiplatform and how it solves code sharing between different platforms.

Chapter 3, Introducing Kotlin for Swift Developers, provides a brief introduction to Kotlin, to bring everyone up to speed before the learning project.

Chapter 4, Introducing the KMM Learning Project, describes the learning project and its prerequisites.

Chapter 5, Writing Shared Code, gives you practical advice on writing shared code with Kotlin Multiplatform.

Chapter 6, Writing the Android Consumer App, explains how the shared code written previously can be consumed on Android.

Chapter 7, Writing an iOS Consumer App, explains how the shared code written previously can be consumed on iOS.

Chapter 8, Exploring Tips and Best Practices, dives deeper into the current state of the art of writing shared code with Kotlin Multiplatform.

Chapter 9, Integrating KMM into Existing Android and iOS Apps, provides tips and answers to possible questions regarding the integration of KMM into existing production apps.

Chapter 10, Summary and Your Next Steps, points you in the next direction in terms of consolidating your KMM knowledge.

To get the most out of this book

You should have basic programming skills, such as familiarity with **Object-Oriented Programming (OOP)**, asynchronous programming, and the general concepts used in mobile development.

Familiarity with Kotlin and Android-related tools (Gradle) is not required as there is a quickstart chapter to bring everyone up to speed on what's needed to start coding the example project.

Software/hardware covered in the book	Operating system requirements
Android Studio Arctic Fox	Windows, Linux, or macOS (to run the project on iOS, you will require macOS)
Android Studio KMM plugin	

If you are using the digital version of this book, we advise you to type the code yourself or access the code from the book's GitHub repository (a link is available in the next section). Doing so will help you avoid any potential errors related to the copying and pasting of code.

Download the example code files

You can download the example code files for this book from GitHub at https://github.com/PacktPublishing/Simplifying-Application-Development-with-Kotlin-Multiplatform-Mobile. If there's an update to the code, it will be updated in the GitHub repository.

We also have other code bundles from our rich catalog of books and videos available at `https://github.com/PacktPublishing/`. Check them out!

Download the color images

We also provide a PDF file that has color images of the screenshots and diagrams used in this book. You can download it here: `https://static.packt-cdn.com/downloads/9781801812580_ColorImages.pdf`.

Conventions used

There are a number of text conventions used throughout this book.

`Code in text`: Indicates code words in the text, database table names, folder names, filenames, file extensions, pathnames, dummy URLs, user input, and Twitter handles. Here is an example: "For every Gradle module/project, you'll have a `build.gradle` (or `.kts`) file describing the build steps for that specific module."

A block of code is set as follows:

```
plugins {
    kotlin("multiplatform")
    id("com.android.library")
    kotlin("plugin.serialization") version
      Versions.KOTLIN_VERSION
    id("org.jetbrains.kotlin.native.cocoapods")
}
```

When we wish to draw your attention to a particular part of a code block, the relevant lines or items are set in bold:

```
// Swift unwrapping
if let string = optional {
    print(string.count)
} else {
    print("I'm nil")
}
```

Any command-line input or output is written as follows:

```
$ mkdir css
$ cd css
```

Bold: Indicates a new term, an important word, or words that you see on screen. For instance, words in menus or dialog boxes appear in **bold**. Here is an example: "Open the Android Studio New Project wizard (**Android Studio | New Project**). From the **Phone and Tablet** tab, select **KMM Application**."

> **Tips or Important Notes**
> Appear like this.

Get in touch

Feedback from our readers is always welcome.

General feedback: If you have questions about any aspect of this book, email us at customercare@packtpub.com and mention the book title in the subject of your message.

Errata: Although we have taken every care to ensure the accuracy of our content, mistakes do happen. If you have found a mistake in this book, we would be grateful if you would report this to us. Please visit www.packtpub.com/support/errata and fill in the form.

Piracy: If you come across any illegal copies of our works in any form on the internet, we would be grateful if you would provide us with the location address or website name. Please contact us at copyright@packt.com with a link to the material.

If you are interested in becoming an author: If there is a topic that you have expertise in and you are interested in either writing or contributing to a book, please visit authors.packtpub.com.

Share Your Thoughts

Once you've read *Simplifying Application Development with Kotlin Multiplatform Mobile*, we'd love to hear your thoughts! Scan the QR code below to go straight to the Amazon review page for this book and share your feedback.

https://packt.link/r/1801812586

Your review is important to us and the tech community and will help us make sure we're delivering excellent quality content.

Section 1 - Getting Started with Multiplatform Mobile Development Using Kotlin

This section covers the core concepts of Kotlin Multiplatform. Comparing cross-platform to native technologies, it describes the market gap that Kotlin Multiplatform fills.

Moreover, it explains the rationale behind why this new technology is one to learn and adopt while giving a deep dive into its architecture and how it makes sharing code between different platforms possible.

This section comprises the following chapters:

1

The Battle Between Native, Cross-Platform, and Multiplatform

The proliferation of smartphones has led to the development of a large number of applications, making app development an important field. Because the same service in the form of an application needs to be developed on multiple platforms, various technologies, in addition to **native** solutions, have started to arise – first, cross-platform and now multiplatform. These technologies have been developed mainly to cut costs and make the application development process more efficient.

We'll start by learning about the issues with native development, why **cross-platform** can solve some of these issues with compromises, and how **multiplatform** can be a better solution. Also, if you're at the start of your career, we'll dive into why it may be better to focus on a multiplatform technology stack, rather than a cross-platform technology stack. I realize that this chapter is quite theoretical, but I encourage you to bear with me – it should help you build up that dopamine release, which is going to be paramount in the following chapters to maximize your learning.

In this chapter, we're going to cover the following topics:

- Understanding the compounding costs of native development
- Exploring the pitfalls of cross-platform solutions
- Adopting a multiplatform approach

Understanding the compounding costs of native development

The manufacturer of every platform or operating system provides a **software development kit** (**SDK**), which contains everything necessary for someone to develop applications on that specific platform or OS. Here, we are referring to a native development process, where someone uses that SDK to develop applications for that single platform.

Cross-platform frameworks have a separate SDK, which is usually a layer on top of a native SDK.

Cross-platform solutions are becoming more and more popular; for example, as of May 2021, out of ~5 million apps on the Google Play Store, more than 200,000 are Flutter-based apps, which is not bad for fairly new technology (4-6% of all the apps published in Google Play Store).

For a more detailed look at some of Google Play's statistics, visit `https://www.appventurez.com/blog/google-play-store-statistics`.

If you are interested in learning a bit more about Flutter-based applications, check out `https://www.youtube.com/watch?v=a553D0s7HeE&t=1779s`.

To understand why there is an increasing demand for cross-platform solutions, we need to understand what issues people face with native development.

One of the reasons for the increase in demand for cross-platform solutions is developer convenience. Becoming an expert nowadays, and especially staying one, in any programming language or framework is not an easy job. While transferring concepts and general knowledge can be achieved in varying degrees, depending on the similarity between two platforms, becoming an expert in a new language still requires learning. Therefore, those people who'd like to become an all-around frontend developer with considerable expertise in Android, the web, and iOS have to learn not only three different frameworks but their primary languages as well: *Swift/Obj-C*, *Kotlin/Java*, and *JavaScript*. This is the case unless there is a shortcut providing passage between these worlds, which is generally covered by cross-platform technologies.

This explains the openness developers have toward using cross-platform, but it's only one part of the equation – the supply – and we still need demand for it.

There is another major reason for developers steering away from native solutions: cost.

The cost of native app development

So, why doesn't everyone want to go with a native development process? This choice is somewhat similar to buying tailored garments versus ready-made garments from clothing stores: it's cheaper.

Before we understand the costs associated with Native development, let's introduce the concept of **nativeness** first. Nativeness is a measure of the degree to which the quality of a product conforms with the peculiarities of a platform. For example, imagine the differences between a native English speaker and a non-native one (potentially the author of this book, who has an imperfect Hungarian accent). The differences can range from subtle to more obvious, based on the complexity of the words and the non-native speaker's skills.

Now, why would anyone give up nativeness? It's mainly because people can achieve lower costs (or at least they think they can).

To get a better picture of the cost variance between native, cross-platform, and multiplatform, we're going to examine the relationship between a feature's complexity and the **development cost** that's needed to bring it to life in a simplistic manner. Features can consist of multiple sub-features. So, for example, a delivery app can be thought of as an app with one delivery feature, where the complexity of this feature is the sum of all of its sub-features.

In the case of native development, since there is little to no cost reduction, the cost of development is determined as follows:

Cost of development (n) = n * FC

Here, *n* is the *number of platforms* and *FC* is the *feature complexity*, which, as we mentioned earlier, is the sum of all the sub-features that comprise a feature.

> **Important Note**
>
> This and the following calculations are approximations and only describe the reality simplistically. Nevertheless, they should provide you with a better picture to understand the cost differences between frameworks.

This is what costs would look like if you were developing a product on two (blue line) and three (red line) platforms, respectively, where there is no cost reduction by sharing code:

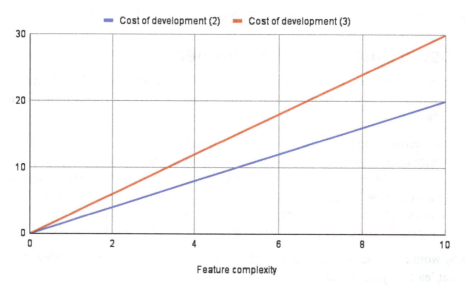

Figure 1.1 – Cost of native development as a function of feature complexity

Unfortunately, there is a little beast known as synchronization between platforms that we didn't take into account, which can significantly increase development costs, bit by bit; it's hard to plan for it, so it can be an unknown variable in calculations.

Synchronization

What is the specialty of frontend and mobile products in general? They are mostly similar, though they do have some differences. Thus, the goal of developers is to achieve consistency between platforms while paying attention to nativeness. This is a lot harder than it sounds. Why? There are a couple of reasons for this:

- People think differently.
- Platforms are different (an option that's easy to implement on iOS may not even be available to Android).
- Creating software requirements documentation that covers everything is impossible.
- Communication is costly, but no communication is costlier.

Because people think differently and are biased, communication is not easy, and platforms can drive developers toward different solutions, platform-native apps will likely have differences. As feature complexity increases, implementations will likely begin to differ more and more, causing greater and greater differences between platforms. Due to this, the costs of synchronization will compound. At some point, the development team will have to account for the differences between the implementations on the platforms as well.

Taking the synchronization costs into account, we could update our cost of development calculation as follows:

Cost of development (n) = n * FC + Sync Costs ^ FC

Here, *n* is the *number of platforms* and *FC* is the *feature complexity*.

Synchronization costs typically depend on your team's processes and its ability to communicate. The following chart provides an example of how synchronization in the native world could increase your costs significantly as feature complexity grows:

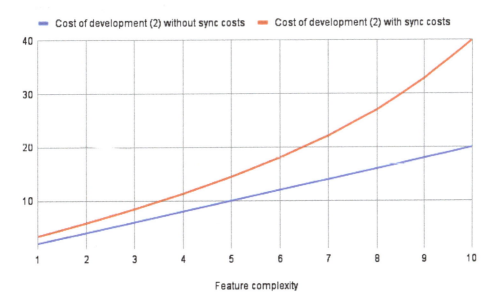

Figure 1.2 – The cost of native development with and without synchronization costs as a function of feature complexity

Here, we can see why the outlook of costs scares clients and directs them toward cross-platform solutions. But do cross-platform technologies save costs? Yes, though not in all cases and they may lure you into traps.

Exploring the pitfalls of cross-platform solutions

The main objective of cross-platform technologies is to allow you to write code that can be used across platforms (Android, iOS, and the web). Due to this, you don't have to write separate code for the same feature multiple times, depending on the platform; the cross-platform framework will provide the tools for you to interpret this code and translate it into platform-specific versions. The power of the cross-platform framework depends on those tools that interpret and translate the cross-platform code into platform-specific code.

Let's learn what the assumed cross-platform development costs are and what you should know about cross-platform in general to avoid some common pitfalls.

Assumed cross-platform development costs

People often estimate cross-platform product costs by cutting the costs that are needed for native apps in half (or even into three, if there is a possibility of deploying the cross-platform app on the web too).

Under this assumption, our formula becomes as follows:

Cost of development (n) = FC

Here, *n* is the *number of platforms* and *FC* is the *feature complexity*.

Let's compare this to the costs of native development:

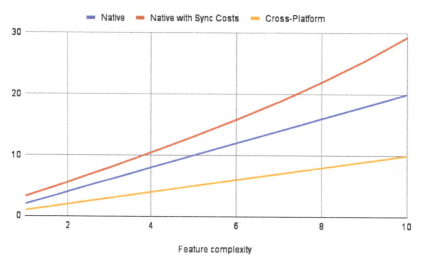

Figure 1.3 – The cost of native development versus the cost of cross-platform development as a function of feature complexity

Looking at the preceding diagram and keeping the aforementioned assumption in mind, no wonder there is an increasing demand for cross-platform solutions.

Though this assumption may hold for greenfield projects, this probably won't be the case for real-world projects. To understand this, let's go over some of the currently available cross-platform technologies and how they work. We will review two of the most popular cross-platform frameworks: **React Native** and **Flutter**.

React Native

React Native is an open source framework for developing mobile applications. It is based on the React library and converts React components and JavaScript code into native Android and iOS components. For example, a `Text` component in React Native will be converted into a `UITextView` component on iOS and a `TextView` component on Android. This sounds like a good approach and it is a plausible one, especially for developers coming from the Web/React world. But how does this conversion work and what are the tradeoffs and risks of React Native development?

React Native creates a thread where it runs the respective JavaScript code, which communicates with the native code by running on the traditional main thread, through a bridge that asynchronously sends serializable data:

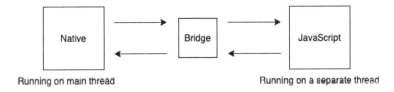

Figure 1.4 – The architecture of React Native

Going back to our example, when a `UIView` or `TextView` is clicked in the native component, the appropriate data is then sent through the bridge to the JavaScript code, and then back again. Now, if you're thinking about the performance costs of this bridge mechanism, then you're in the right place. Let's look at the drawbacks of React Native:

- **Performance**: It's not native, especially for resource-intensive features.
- **New features support**: Because you're relying on React Native to provide support for new things, you can expect a bit of a delay.

There are also some application development specifics, such as permissions, notifications, in-app purchases, and media where you'd like more control over the native platform's API. In those cases, React Native lets you create native modules in regular native code, though it's not the primary purpose of the framework. If you arrive at a point where you need a native module, which is likely unless you have a really simple app, you will face the following issues:

- **As a Developer**: If you planned to reuse your JavaScript and/or React knowledge to create mobile applications, you will have to acquire native mobile development skills anyway.

- **As a Client**: Every roadblock that pushes you toward implementing a native module means higher costs than writing the same feature with native solutions, simply because there is a need for native expertise. Plus, it has to be integrated with React Native as well.

We'll update our charts and calculations in a moment, but first, let's check out Flutter.

Flutter

Flutter is an open source UI software development kit that's developed by Google and used for developing cross-platform applications. It has three layers from an architectural perspective – the framework, the engine, and the platform – and relies on Dart's language specifics, such as ahead-of-time compilation.

As a developer, you interact with the framework and you write the app and the widgets (UI components in Flutter) in a declarative way using Dart, which the engine then renders to a canvas called Skia Canvas. This canvas is then sent to the native platforms: Android, iOS, or the web. The native platform will show the canvas and send the occurring events back:

Figure 1.5 – The architecture of Flutter

Flutter's architecture may be similar to React Native, but there is a big difference in terms of performance. One key component of how Flutter achieves better performance than React Native is by going one level lower on the native side, meaning that it doesn't use the traditional SDKs that are used by native developers. Instead, it uses SDKs that need more developer expertise and can offer higher performance. Flutter uses Android's **Native Development Kit** (NDK) and iOS's **Low-Level Virtual Machine** (LLVM) to compile the C/C++ code coming from the engine.

While Flutter has pretty good performance compared to native and is far better than React Native when it comes to compiling the Dart code into a lower level native code (a key performance component), it also has a drawback: the cost of writing native code with Flutter is higher than using React Native to do the same.

At the time of writing, if you don't have support for a certain piece of functionality in the Flutter framework itself, you can write regular Java/Kotlin and Obj-C/Swift code, but you'll have to communicate with the Dart code through a channel, sending data through `Map` in Dart, `HashMap` in Java/Kotlin, and `Dictionary` in Swift. If we compare this to the regular **Java <-> Kotlin** or **Obj-C <-> Swift** interoperability, this can be perceived more as a workaround than a scalable solution.

> **Important Note**
>
> Both the Flutter and the React Native descriptions only serve as high-level overviews to help you understand how cross-platform solutions are designed and what to expect when you're working with them. To get a clearer picture, please read the official documentation.

To conclude our cross-platform overview, let's summarize the patterns that we observed in the aforementioned frameworks and see how we can update the general assumption of cross-platform solutions when it comes to estimating the costs of development.

The main ideology of cross-platform technologies is that you write the same code for Android and iOS (and the web); the framework provides the tooling to interpret this code and translate it into the platform-specific version.

While they do provide solutions for writing native code where needed, they are suboptimal and the goal of any cross-platform project is to avoid situations where interoperability with native code is needed.

This way, you rely heavily on the framework to make good decisions on your behalf when you're translating the cross-platform code into the platform-specific version. In short, all of these frameworks have, or will have, a tough time keeping up-to-date with both Android and iOS, two platforms that don't have an incentive to stay in sync with each other.

So, unless you plan on accepting big compromises, your cost of maintaining an acceptable level of nativeness will be relatively high with any cross-platform solution.

Actual cross-platform technology costs

Going back to our initial cross-platform costs assumption, we can update our formula with a new variable:

*Cost of development (n) = FC * (1 + Cost of going Native)*

Here, *n* is the *number of platforms* and *FC* is the *feature complexity*.

The *Cost of going Native* can depend on a variety of things:

- How much interoperability the cross-platform technology provides with native. We've seen that this isn't optimal with neither of the aforementioned technologies.

- The knowledge gap between the cross-platform and native languages. You'll likely observe that expertise hardly translates from cross-platform to native.

- The more you need to dive into native implementations, the more your costs will compound because synchronization costs will kick in for the native code as well.

For visualization purposes, a more likely scenario of the costs associated with cross-platform development could look like this:

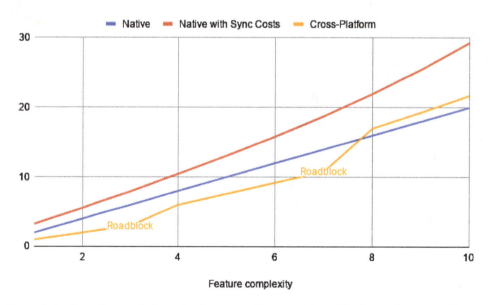

Figure 1.6 – Cost of cross-platform development with potential roadblocks as a function of feature complexity

Again, the number of roadblocks you'll hit heavily depends on how much you're willing to compromise from nativeness and how much you're relying on platform-specific APIs.

To conclude, if I were to write a project for myself, I'd consider Flutter. If it is a simple project where I don't have to cover any platform-dependent stuff (permissions, notifications, in-app purchases), just basic CRUD operations with a backend, a local database, and some nice UI stuff, then I'd probably go with Flutter. Otherwise, I'd use a native solution. Knowing how platform-specific things such as permission handling change on Android, I wouldn't dare trust a third-party framework to keep up-to-date.

That being said, cross-platform will probably still attract many start-ups in the future, due to the nature of start-ups accepting higher compromises to survive and achieve their short-term financial goals or to be product-market fit, which requires moving fast. However, there is another option: the multiplatform approach. This is cost-friendly both long and short term, and it is a sane approach from all perspectives.

Adopting a multiplatform approach

We have finally arrived at one of my favorite topics. In this section, we'll explore how multiplatform works, why it's different from cross-platform technologies, and its cost implications.

The multiplatform approach

As we mentioned previously, cross-platform technologies generally try to take on the "burden" of dealing with platform-specifics; thus, their main goal is to help facilitate application development without having to deal with platform-specific decisions. This has two implications – interoperability with the native platform is not the primary scope of these technologies and (partly because of this) the framework needs to do most of the heavy lifting when it comes to making platform-specific decisions.

To overcome these issues, another approach is needed. **Kotlin Multiplatform (KMP)**, a multiplatform solution, introduces a paradigm shift. It recognizes both the need for flexible platform-specific decision making and keeping up to date with different platforms, where these two things go hand-in-hand.

Its aim is not to provide a wrapper layer over the native platforms, but to be a handy tool in the native development palette, which can help with sharing non-platform-specific code such as the business logic.

You may be wondering why understanding the ideology of a framework would be important for you. There are a couple of reasons, as follows:

- You become more aligned with a framework, and you'll know when something goes against the framework's design.
- You'll be able to manage your expectations regarding the framework's future direction better.

The main objectives of KMP are as follows:

- Keeping the native part of development as close to the regular native development process as possible.
- Ensuring that native developers do not find it difficult when they're writing the shared code.
- Facilitating interoperability between native and shared code; interacting with shared code should be as close to native-like as possible.

Now, let's take a deeper look into how KMP can empower you to write platform-agnostic code and share that between different platforms.

How KMP works

KMP allows you to write code in Kotlin in a platform-agnostic way and share that code between different platforms, all while leveraging the native programming benefits.

The Kotlin ecosystem contains three main compilers – Kotlin/JVM, Kotlin/JS, and Kotlin/Native (we will cover them in more depth in *Chapter 2, Exploring the Three Compilers of Kotlin Multiplatform*):

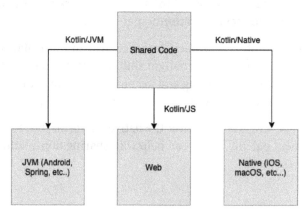

Figure 1.7 – The architecture of KMP

> **Note**
>
> Most developers know Kotlin through the lens of Kotlin/JVM. This is because of Kotlin's reliable interoperability with Java. It has a wide and quickly growing adoption rate in the Android community, but server-side development with Kotlin has also been becoming more and more popular in recent years.

In essence, Kotlin's interoperability power depends on how well these three compilers work with the respective platforms. For Android, we can consider that the interoperability cost with KMP is zero since Kotlin/JVM is part of the Android developer ecosystem. As for iOS (and potentially the web), the costs depend on how well the Kotlin/Native (and Kotlin/JS) compiler works. We will look at this in more detail in the next chapter.

Shared code must be platform-agnostic, which means the code shouldn't contain any JVM, JavaScript, iOS, or any other platform-specific references. For example, working with `Date` and `Time` is platform-specific and has different dependencies on iOS than on JVM (or Android). Don't worry – a lot of these use cases are already covered in libraries that have been developed by either the Kotlin community or the JetBrains team.

Now, let's learn how to leverage KMP's capabilities to write platform-agnostic code that will use the proper platform-specific dependencies on the different target platforms, in case you bump into any uncovered use case from the community.

Platform abstractions (expect-actual)

This mechanism is one of the cores of the whole KMP technology. In a lot of cases, when you're writing shared code, you need a way to define how certain functionality should be implemented on the specific native platforms.

> **Note**
>
> Going forward, we will use the terms **shared code** and **common code** interchangeably, both of which refer to code written in a platform-agnostic way and that could be seamlessly compiled with one of the Kotlin compilers to the chosen targets.
>
> As you'll see, this could mean being platform-agnostic across Kotlin/Native and Kotlin/JVM only, depending on what platforms you target.

With KMP, you can write **expected declarations** using the `expect` keyword in your shared code, which will have an `actual` implementation for every platform that you specify. Let's look at an example of how to share code between Android and iOS with this mechanism.

Let's say you have an application where users can upload certain files to the cloud and you'd like to share this part of your networking layer. Since file handling is something platform-specific, you'll need to create some abstractions for this (or potentially check if it's already covered in a library).

First, you would declare the *expected* functionality in your shared code; in our case, we'll need any file to be converted into a byte representation of the file so that we can send it to the backend:

```
expect class File {
    fun toByteArray(): ByteArray
}
```

Don't worry much about the syntax; the important part is the expect/actual mechanism.

To make KMP able to substitute the expected implementations with the actual implementations on the different platforms, we need to provide those as well:

```
// JVM
actual class File(private val file: java.io.File) {
    actual fun toByteArray() = file.readBytes()
}
```

As you can see, for JVM/Android, we are just wrapping the java.io.File platform-specific implementation. There is a better way to do this using type aliases, which we'll cover in *Chapter 5, Writing Shared Code.*

For iOS/React Native, the implementation could look like this:

```
// iOS/Native
actual class File(private val fileHandle:
 platform.Foundation.NSFileHandle) {

actual fun toByteArray() =
  with(fileHandle.readDataToEndOfFile()) {
      memScoped {
          ByteArray(length.toInt()).apply {
              usePinned {
                  memcpy(it.addressOf(0), bytes, length)
              }
```

```
                }
            }
        }
    }
```

As you can see, in the native implementation, you can use the **Foundation Kit**; there is also a way to include CocoaPods dependencies, which we will also cover in *Chapter 5, Writing Shared Code.*

At this point, KMP will compile the shared code for two different targets (Kotlin/JVM and Kotlin/Native) with the two different compilers and replace all the expected declarations with their `actual` implementations on the specific platform.

I can't emphasize enough how important this mechanism is for multiplatform; this is what enables the bridge between different platforms and provides the scalability for the whole platform so that outside contributors can easily build upon the current solutions.

Next, I'm going to touch on a little tool that we're going to cover in more depth in *Chapter 2, Exploring the Three Compilers of Kotlin Multiplatform*, which helps out tremendously with actual implementations – the **commonizer**.

This tool automates the process of the expect/actual declaration and generates the expect/actual declarations for us. However, this tool was designed specifically for cases where targets (such as macOS and different iOS architectures) have very similar dependencies (such as the POSIX library on OS X and Linux).

Now that we have a bit of an understanding of the KMP framework and how it enables developers to share code, let's see what it can be used for and how it could help out in a regular development process.

The different use cases for KMP

The KMP framework is unopinionated about what you use it for. Its main goal is to help you share code between multiple target platforms with as good interoperability as possible.

This means that the possible combination of potential use cases is close to infinite. You can play around with the amount of code you plan on sharing and the targets you'd like to share the code between. You can also scale it later on in the process because you can add other target platforms and migrate more and more code to your common part as your project develops.

You can go from having 1% shared code to sharing your UI layer – the only blocking thing will be your sense of what needs to stay platform-specific.

With this in mind, let's check out some of the most common use cases.

Kotlin Multiplatform Mobile (KMM)

You may have heard about **Kotlin Multiplatform Mobile** (**KMM**) and perhaps you've been wondering what the difference is between KMP and KMM; allow me to shed a bit of light on this topic.

Technology-wise, KMM is a specific use case, whereas KMP is used for sharing code between mobile targets – Android and iOS.

KMM was introduced when JetBrains realized that this concept is, at the time of writing, one of the main use cases for developers choosing KMP to share code. Hence, a dedicated KMM team was formed and special tooling was introduced to help support this cause:

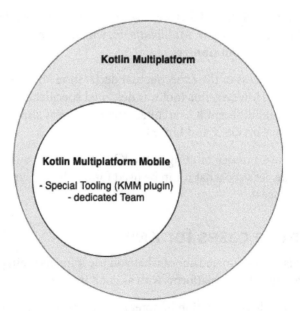

Figure 1.8 – Kotlin Multiplatform Mobile in the Kotlin Multiplatform technology

In KMM, your code-sharing capabilities will largely depend on two of the Kotlin compilers: Kotlin/JVM and Kotlin/Native. To grasp the limits of what's capable when working with these compilers, we dedicate *Chapter 2, Exploring the Three Compilers of Kotlin Multiplatform* to this so that you can know what to expect and how to get the most out of both the Kotlin/JVM and Kotlin/Native compilers.

As we've already mentioned, you can start with any level of code sharing, but here are some examples:

- **A Small Part of the Code Base**: Kevin Galligan would say to choose one of the parts that's not so fun to work on, such as analytics.

- **Networking Layer or Persistence Layer**: This is still a relatively small amount of the code base and it can reduce some of the synchronization costs.

- **The Entire Data Layer**: Managing offline support and syncing logic consistently on two different platforms can be a burden, so it can be worth doing this for certain apps.

- **View/Presentation Layer**: This can be done, but things get a bit more platform-specific here. This is also where the line between cross-platform and multiplatform starts to get a bit blurry.

You can start going from only a small part of the code base and then bring more and more layers and/or features as you gain more confidence working with KMP.

Another nice benefit of KMM is that it doesn't change the native development cycle radically. Instead, it builds upon it, with KMP being more of an additional tool in the existing palette.

Going forward, this use case is going to be the main focus of this book, but we will briefly explore other potential use cases so that you can get a better picture of what code-sharing possibilities you have with KMP.

Code sharing between frontend applications

You can do this gradually as well, going from a KMM app to sharing logic between all the different frontend platforms you plan on supporting.

Since your current shared code is already based on working with the Kotlin/JVM and Kotlin/Native compilers, adding support for all the different desktop targets such as macOS, Windows, and Linux is relatively easy and largely depends on how well you manage the non-shared part of your code.

A slightly bigger step is to bring the Kotlin/JS compiler into play and share code with your web app through a JS target.

The complexity of this depends on the interoperability power of Kotlin/JS and how well you can work with it.

Code sharing between backend and frontend applications

Another interesting use case of KMP is sharing code between your backend and frontend applications.

In most real-world projects, there is a limited amount of implementation overlap between backend and frontend apps, so this is why it doesn't get much focus from cross-platform solutions.

Nevertheless, there is always a piece of the backend that would be awesome to share. I've had the chance to experience minor modifications that broke the frontend apps, and also remember doing Git history research to understand why there are differences in the way frontend platforms use the backend APIs.

Yes, you can minimize these human errors with carefully designed processes, but enforcing the process itself can be another challenge.

I think that sharing DTOs, API keys, and other useful information, such as base URLs, can speed up development, especially in the long term. Just think about a **continuous integration** (**CI**) pipeline, where if a backend modification breaks the builds on the apps, it's immediately visible to the backend team.

I think that the combination of use cases is huge, and as a developer, I would start getting more and more into this world that KMP offers. The whole approach offers a new perspective on how we think about developing apps and introduces a new potential team composition:

- **Platform Experts**: Developers with native Android, iOS, web, or other platform expertise
- **Shared Code Experts**: The ones who maintain the shared logic and know the ins and outs of KMP

JetBrains had already started experimenting with this setup while developing their Space product and as KMP expertise spreads, I suspect we will see even more people follow.

Now, let's close this chapter by talking about the cost implications of a multiplatform approach.

KMM cost implications

At this point, you hopefully understand the differences between cross-platform, native, and multiplatform. The latter is in-between a native and cross-platform solution, where you remain with your native platform development cycle but enhance it with code sharing capabilities where it makes sense to.

So, how would you calculate the costs for a multiplatform project? It should have native costs for your non-shared code and cross-platform costs for shared code, except that you don't face roadblocks with KMP, with interoperability being much better than it is with cross-platform solutions.

In the case of a real roadblock, you can just decide on not sharing that part of the code so that your interop costs will be diminishing relative to those of cross-platform solutions.

Based on this reasoning, a possible calculation of KMP costs could look like this:

Cost of development (n) = FC * [n * (1 - α) + α]

Here, *n* is the *number of platforms*, *FC* is the *feature complexity*, and α ∈ [0,1] represents the *amount of shared code* (1: all the code is shared, 0: no code is shared).

Note that in this case, we don't include any synchronization costs. This is because KMP, when done right, should eliminate the situations where synchronization costs could occur; thus, the non-shared amount of code should be a representation of the platform-specific code that's not worth sharing.

Of course, since KMP is a relatively newborn platform, the aforementioned ideal scenario probably won't manifest for every use case, though it is approachable. To grasp what this cost calculation means, check out the following chart:

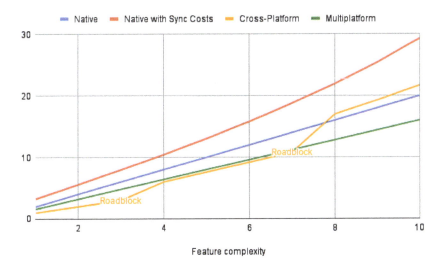

Figure 1.9 – Cost of KMP development versus other options as a function of feature complexity

As you can see, as the costs of development increase, cross-platform solutions can be a good choice for short-term, quick projects. But in the long term, KMP is going to be the winner.

> **Important Note**
>
> I'm going to emphasize again that this is a simplistic estimation of costs and that the preceding chart is a representation of a fabricated (though possible) scenario of project development.
>
> Because estimating real-world projects with generic calculation logic and from the perspective of the different technologies is a hugely complex task, this should be enough reasoning as to why this simplistic approach was taken.
>
> Nevertheless, I'm confident that this simplistic approach can provide a good overview of the costs of the different technologies.

Please note that an important aspect is missing from this chart – having an even better view that shows another dimension of the quality would be required to have complete reasoning on the technologies.

We won't dive deeper into this topic, but I'd reason about my technology choices in the following manner:

- Are quality and nativeness paramount for my project? If the answer is yes, go as native as possible.

- If both quality and costs are important and you're looking for the highest quality/ cost ratio, then go with multiplatform. Note that KMP is applicable for the first scenario as well since it offers gradual code sharing; hence, if you only find out that sharing something affects your quality during the process, you can revert and go fully native for that feature. The upside is that you'll cut a lot of the costs.

- Cross-platform is the most cost-efficient option, but it is likely to require compromises.

Summary

At this point, you should have a better understanding of the different technologies for mobile development and their cost-effectiveness. I also hope that you've become eager to learn more about Kotlin and KMP.

The purpose of this chapter was to provide a good overview of why KMP is different from other cross-platform technologies, why it can be a good career choice for developers, and why it would make sense to choose it for a project.

Now that you've had that dopamine pump, let's dive into the more technical aspects and check out how the Kotlin compilers work and how you can leverage their power.

2

Exploring the Three Compilers of Kotlin Multiplatform

In the previous chapter, we discussed that the interoperability quality of shared code is a key aspect of multiplatform development. To explore this interoperability quality, we need to examine how the three different backend compilers of Kotlin – **Kotlin/JVM**, **Kotlin/Native**, and **Kotlin/JS** – work. This will help you manage your expectations regarding the performance, future, and interoperability of Kotlin with the different platforms, which will help you leverage the potential of KMP.

By the end of this chapter, you will have a clearer picture of how the aforementioned compilers work, what interoperability constraints you'll have when working with them, and how to leverage their power.

In this chapter, we're going to cover the following topics:

- Kotlin compilers in general
- The Kotlin/JVM compiler
- The Kotlin/Native compiler
- The Kotlin/JS compiler

Kotlin compilers in general

First, let's make sure we are on the same page and have a basic understanding of how compilers and the Kotlin compiler work in general.

A compiler is a program that translates computer code in a given programming language into machine code or lower-level code. Compilers generally consist of two components:

- Frontend
- Backend

A **frontend compiler** deals with programming-language specifics, such as parsing the code, verifying syntax and semantic correctness, type checking, and building up the syntax tree. Generally, there is one frontend compiler and as many backend compilers as there are targets.

A **backend compiler** takes an **intermediate representation** (**IR**) of the code that's produced by the frontend compiler and creates an executable based on the IR. This can be run on the specific target while running certain optimizations.

In Kotlin, there are three different backend compilers: one for each of the **Java virtual machine** (**JVM**), **JavaScript** (**JS**), and native targets. All three produce different outputs that will conform to the target platform.

Until now, the three Kotlin backend compilers were developed pretty much independently, without much overlap. JetBrains recently started a new direction by introducing an IR for Kotlin code, which is already adopted in Kotlin/Native.

The Kotlin/JS and Kotlin/JVM backend compilers are being migrated to this new IR infrastructure at the time of writing; hopefully, they will have more stable versions once this book has been published.

This new unified IR-based compiler means that all three backend compilers will share the same logic, thus making feature development and bug fixes easier. It also brings the possibility of multiplatform compiler extensions, which could be pretty neat.

Now, let's look at the different Kotlin backend compilers.

The Kotlin/JVM compiler

The Kotlin/JVM backend compiler is what helps translate code written in Kotlin into Java bytecode, which is code that can be run on the JVM or Android. Kotlin was initially designed for the Java world, including Android, and the Kotlin/JVM compiler was the one that paved the way for the Kotlin language.

How it works

The Kotlin/JVM compiler generates the same `.class` executables that the Java compiler does, which is the Java bytecode that can be run on the JVM:

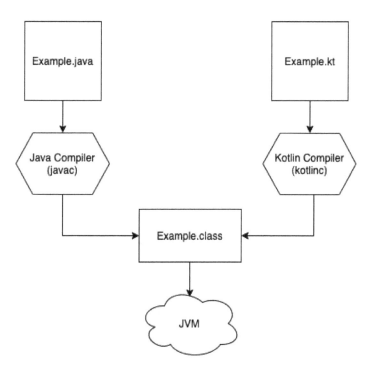

Figure 2.1 – How Kotlin/JVM works

This means that you can decompile your Java bytecode, the `.class` executables, and check the Java code, which is quite handy if you want to see what the generated Kotlin code looks like.

So, the Kotlin/JVM value proposition was (and still is) that it provides the rich palette of language features of Kotlin and translates the code you write with it into the same Java bytecode that has seamless interoperability with any other Java code.

This strong interoperability feature of Kotlin/JVM has led to a rapid rise in the number of people trusting Kotlin and choosing it over Java when developing Android apps. This huge community growth and the official Google support for Kotlin have evolved to a point where many of the official Android libraries are Kotlin-first.

Android is now moving to a new UI toolkit, **Jetpack Compose**, where the underlying Compose compiler completely relies on the Kotlin compiler. All this means that the Google team is now even more invested in Kotlin, which can be seen in their contribution to the Kotlin/JVM compiler infrastructure as well. The Compose compiler uses a newly introduced infrastructure of the Kotlin/JVM backend compiler.

See the following talk for more information: `https://www.youtube.com/watch?v=UryyHq45Y_8`.

This means that the Kotlin/JVM backend compiler is currently the most supported compiler by JetBrains, Google, and the huge Android community. Though things have been changing for the best, since Kotlin 1.5, the new unified IR-based Kotlin/JVM backend compiler became stable and enabled by default. This means that the other backend compilers can potentially benefit from any feature and bug fixes on the Kotlin/JVM compiler.

I also have to mention that the big success of Kotlin/JVM on Android was probably also helped by the way Android runs Java executables.

Executing Java code on Android

Applications running on mobile phones have more constraints and fewer resources than applications running on server and desktop environments. Using a VM not only helped Android support the vast number of hardware but in part also optimized the mobile environment. Dan Bornstein designed the **Dalvik Virtual Machine** (**DVM**), which is based on JVM and is specifically for Android devices.

This means that running the `.class` Java bytecode on Android isn't exactly as straightforward, because there isn't a JVM. This bytecode needs to be translated by the DVM; this is what the **d8** Dex compiler does – it takes the Java bytecode and produces Dalvik bytecode or `.dex`:

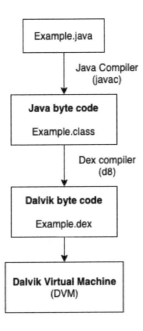

Figure 2.2 – How the DVM works

So, Android needs to support any new Java versions in their Dex compiler before a developer can use the features of the new Java release.

The key takeaway is that Android developers have to wait for Dex to have support for that version because of how Java runs on Android.

This means that there's breathing time for Kotlin because until the Dex compiler doesn't support the new Java release, there is no point in supporting it in Kotlin, at least for Android developers.

Note that Android replaced DVM with **Android Runtime** (**ART**). DVM used **just-in-time** (**JIT**) compilation, which means that every app was compiled before it was launched. ART introduced **ahead-of-time** (**AOT**) compilation, which, during the application's installation phase, takes the Dalvik bytecode, translates it into the machine code, and stores it. Note that ART still includes a JIT compiler, which complements the AOT compiler by continually improving the performance of Android applications as they run. See `https://source.android.com/devices/tech/dalvik/jit-compiler?hl=en` and `https://source.android.com/devices/tech/dalvik#AOT_compilation` for more information.

Now that we've finished this small Android detour, you should have a good understanding of how the Kotlin/JVM backend compiler works, how well it is supported, and the enabling factors that led to its success.

Now, let's dive into the Kotlin/Native compiler, understand how it works, and how you can leverage it when you're trying to share code with iOS and other targets from the Apple ecosystem.

The Kotlin/Native compiler

The Kotlin/Native backend compiler is an LLVM-based compiler (the abbreviation stands for *low-level virtual machine*, which was officially deprecated to avoid any confusion since LLVM now means more than just a **virtual machine** (**VM**); we're talking about LLVM IR, LLVM debugger, and so on) that compiles Kotlin code into native binaries that can be run without a VM. It can be used to compile code for embedded devices, the Android **Native Development Kit** (**NDK**) or iOS, macOS, and other Apple targets.

We can immediately draw some comparisons here with Flutter, which uses the Android NDK and LLVM to compile Dart on Android and iOS, respectively; this is known to be one of the key factors of Flutter's pretty good performance compared to React Native.

One of Kotlin/Native's powers comes from the fact that it can provide complete two-way interoperability with the Native targets. This means that you can use the C, Swift, and Objective-C frameworks and static or dynamic C libraries in your shared Kotlin code (we saw this in *Chapter 1*, *The Battle Between Native, Cross-Platform, and Multiplatform*, where we wrote an actual file implementation based on `NSFileHandle`).

> **Note**
>
> The Kotlin/Native compiler can create an executable for many platforms, a static library or dynamic library with C headers for C/C++ projects, and an Apple framework for Swift and Objective-C projects.

How it works

First, the Kotlin/Native compiler generates an LLVM IR of the original Kotlin code. Then, the LLVM compiler can work with this IR and create the necessary executables. This includes binaries or frameworks in the case of the Apple ecosystem:

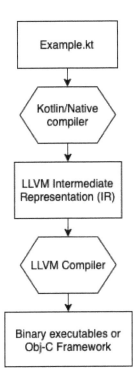

Figure 2.3 – How Kotlin/Native works

Interoperability on iOS

Let's look at what the experience of consuming shared code written in Kotlin looks like on iOS.

With Kotlin/Native, you can generate not only binary executables but Obj-C frameworks for Apple targets. The way this works is that Kotlin/Native compiles Kotlin directly into native code with the help of an LLVM and generates some adapters/bridges to make this compiled Kotlin code accessible from Obj-C and Swift.

This means that if you're using Swift (which you most likely are), you have interoperability that looks like this: **Kotlin <-> Obj-C <-> Swift**.

This means that Obj-C acts as sort of a bridge between Kotlin and Swift, so to use a Swift library in Kotlin, the given library must be usable in Obj-C as well. This can be done by exporting the Swift library's API to Obj-C via `@objc` annotations.

Pure Swift modules, without these annotations, cannot be used, which means that you cannot base your actual implementations on modules such as `SwiftUI`, for example.

> **Note**
>
> Note that JetBrains has added direct interoperability with Swift to their roadmap, but its development is currently paused.
>
> This means that the Kotlin/Native compiler will not generate an Obj-C-specific adapter for the native code, but rather a Swift one.

To get an all-around understanding of what you'll get when you compile your Kotlin code for an Obj-C framework with Kotlin/Native, the best way is to just compile your code and check out what its Obj-C representation looks like. The second best way is to check out this concise one-pager from the official documentation: `https://kotlinlang.org/docs/native-objc-interop.html`.

But to get a better idea of what your Kotlin code could look like on Obj-C, let's look at a relatively common example of a `data class`. Don't worry if you don't know about the concept of a `data class`; in *Chapter 3, Introducing Kotlin for Swift Developers*, we'll explore the core features of the Kotlin language.

Let's go with the following example `data class`:

```
// Kotlin code
data class Example(val param1: String, val param2: String)
```

This class will be compiled to the following Obj-C code:

```
// The generated Obj-C code
__attribute__((objc_subclassing_restricted))
__attribute__((swift_name("Example")))
@interface KotlinIos2Example : KotlinBase
- (instancetype)initWithParam11:(NSString *)param1
    param2:(NSString *)param2
    __attribute__((swift_name("init(param1:param2:)")))
    __attribute__((objc_designated_initializer));
```

```
- (NSString *)component1
  __attribute__((swift_name("component1()")));
- (NSString *)component2
  __attribute__((swift_name("component2()")));
- (KotlinIos2Example *)doCopyParam1:(NSString *)param1
  param2:(NSString *)param2
  __attribute__((swift_name("doCopy(param1:param2:)")));
- (BOOL)isEqual:(id _Nullable)other
  __attribute__((swift_name("isEqual(_:)")));
- (NSUInteger)hash __attribute__((swift_name("hash()")));
- (NSString *)description
  __attribute__((swift_name("description()")));
@property (readonly) NSString *param1
  __attribute__((swift_name("param1")));
@property NSString *param2
  __attribute__((swift_name("param2")));
@end;
```

It may seem that the generated Obj-C code has a lot of additional code compared to the initial Kotlin code, but essentially, what you get with Kotlin data classes is that you can automatically generate the `equals`, `copy`, and `hash` functions.

> **Note**
>
> The preceding Obj-C code snippet, which was generated by the Kotlin/Native compiler, is only a snapshot of the compiler's functionality at the time of writing. As the compiler evolves, the generated code may look different in future versions.

In mobile applications, efficient asynchronous programming is an important factor, so there is a huge demand for high-quality language support to ease this. The Kotlin/JVM compiler does a pretty good job of complying with the Java concurrency model.

Now, let's explore how Kotlin/Native approaches this problem.

The concurrency model

One of the biggest headaches developers have to face in the Kotlin/Native world is bumping into concurrency-related issues in their code and experiencing the strictness of the current concurrency model of Kotlin/Native. Let's quickly cover why these restrictions were introduced and why the JetBrains team is currently working on some changes.

Long story short, because of how the current automatic memory management works on Apple-based native systems (iOS, macOS, and so on), some concurrency restrictions needed to be imposed in the Kotlin/Native world. This means that while mobile developers are pretty much used to being able to share objects between threads freely and have adopted various best practices and patterns to avoid race conditions when doing so, they still have to face the really strict world of Kotlin/Native if they plan on sharing logic that involves concurrency and sharing states across multiple threads.

It is possible to write efficient mobile apps even with these restrictions, but it requires a higher level of expertise and the current model is not perfect; in some edge cases, it introduces certain memory leaks.

All this has slowed down the adoption of KMP, and KMM in particular, and pushed JetBrains toward a new solution. They've announced a new memory management infrastructure, which should enable a more performant and developer-friendly concurrency model in Kotlin/Native.

Nonetheless, the current model will still be supported, and it is necessary to understand it until the new concurrency model arrives at a stable version. We'll start learning about the current concurrency model next, but I hope that after 1 or 2 years of writing this book, all the information in this section will be outdated and that people won't need to spend as much time understanding this topic.

The current state and concurrency model

The strictness of Kotlin/Native's concurrency and state model consists of the following two main rules, where the second is connected to the first:

- Mutable states can't be shared between threads.
- States need to be made immutable to be shared between threads. Their value, since they are immutable, cannot be changed afterward.

KMP shared code, as discussed previously, can be used across multiple platforms from the JVM, Native, and JS worlds. This means that the Kotlin/Native concurrency model will have to be enforced in that specific target only (later in this book, we'll see that this will affect how you design your shared code as well), and in practice, Kotlin/Native's concurrency rules will be enforced at runtime.

So, for the concept of immutability, Kotlin/Native introduces **frozen** objects:

Frozen = Immutable

In code, this means that you need to make all the states that you want to share between threads immutable/frozen. Freezing an object can be done in Kotlin/Native using the `freeze()` function.

> **Important note**
>
> Using `freeze()` not only freezes/makes the object itself immutable, but the whole subgraph, including any other object that can be reached from the object where `freeze()` is called.
>
> Freezing is a one-way ticket, so you can't unfreeze any object.
>
> A common source of crashes in Kotlin/Native applications is accidentally freezing objects that weren't purposefully targeted as freezable objects.

So far, we've discussed that Kotlin/Native enforces the aforementioned rules at runtime; this means that abusing those rules will result in runtime exceptions. Let's look at what exceptions you're likely to bump into in a multithreaded application.

IncorrectDereferenceException

This is the result of the first rule: *mutable state cannot be shared across threads*. Whenever you get an `IncorrectDereferenceException` exception, this means that an object that is unfrozen/mutable is shared across threads.

In practice, this can come up in different scenarios, such as calling a Kotlin function from Obj-C/Swift that runs on a background thread, with parameters created on another thread in Obj-C/Swift, or running shared code that was tested on the main thread only.

InvalidMutabilityException

As its name suggests, this exception is the result of the second rule: *the value of an immutable object cannot be changed*. This will happen any time you're trying to mutate an immutable. In other words, changing the value of a frozen object will cause an `InvalidMutabilityException` exception at runtime.

Unfortunately, in some cases, this is not that easy to debug because, as we mentioned previously, `freeze()` freezes the object itself and any other objects that the target object touches, which means you'll need to find out how or where a certain object was frozen.

A good practice is to implement the `ensureNeverFrozen()` function on objects that you're confident should not be shared across threads. An exception will be thrown right away if the object is already frozen or later on when an attempt to freeze is made.

That is enough of the basics. Now, let's touch on some additional best practices that you can apply during development.

Making sure that developers have a unified and consistent development experience when working with KMP isn't easy because of some of the different features in the JVM, Native, and JS worlds. Some examples include reflection, the state and concurrency model, the memory model, and annotation processing.

Annotation processing and reflection are two other big topics that we need to address since a lot of the libraries that are available in the Kotlin/JVM ecosystem are dependent on them. Having a Native version of them could unlock those libraries and bring their capabilities to the KMP world.

Annotation processing

In Java, annotations became popular because a lot of code generation tools can be created using the Java annotation processor to get rid of boilerplate code. These tools include libraries such as Dagger's **dependency injection** (**DI**), developed by Google, or Room, the persistence library on Android.

In the Kotlin/JVM world, this is done via the **Kotlin Annotation Processing Tool** (**KAPT**), which has two huge drawbacks:

- It has slow performance since KAPT needs to generate intermediate Java stubs, which can then be processed by the Java annotation processor.
- Since it relies on the Java annotation processor, it's impossible to build platform-agnostic libraries on top of KAPT, which could otherwise be used in KMP projects.

But the good news is that there is already a powerful alternative to KAPT. **Kotlin Symbol Processing** (**KSP**) is a new tool developed by the Google folks that offers similar functionality to KAPT. However, since it doesn't rely on Java annotations and the annotation processor, it can offer direct access to Kotlin compiler features, it's multiplatform friendly, and it's up to two times faster than KAPT.

Since Room and Moshi already provide experimental support for KSP, Android developers may have a good chance of not having to give up their current well-tried library choices if they want to build Kotlin Multiplatform apps.

In *Chapter 1, The Battle Between Native, Cross-Platform, and Multiplatform*, we briefly looked at the Commonizer tool. Let's dive deeper into that topic and see what problems it solves and why it was needed.

Intermediate source sets and the Commonizer

KMP's goal is to make code sharing across all targets as convenient as possible. On iOS, there are different CPU architectures. KMP supports the following targets:

- **Arm64**: This is a 64-bit ARM CPU architecture that's supported on iOS 7+. All devices from iOS 11 onward use this architecture.

- **Arm32**: This was used before Arm64.

- **x64**: This is a 64-bit Intel processor that's available for simulators.

This means that if you're developing a KMM application, you probably want to target both the Arm64 and x64 CPU architectures; you don't want to duplicate your actual platform implementations for these targets.

Additionally, if you plan on supporting macOSX64 targets later, you may have some logic between the Apple targets that relies on the same foundation dependencies and could be shared.

In these cases, you need an intermediate source set, that is, an iOS for combining the two Arm64 and x64 targets and Apple for combining the iOS targets with the macOSX64 target:

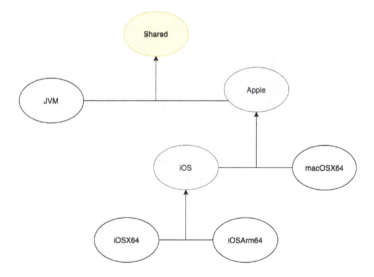

Figure 2.4 – iOS and Apple intermediate source sets

As you can see, you should use the targets provided by the KMP framework and combine them flexibly. By doing this, you would create both iOS and Apple intermediate source sets and define common platform functionality for the specific target set:

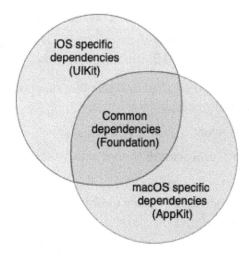

Figure 2.5 – Intermediate source set dependencies

At first, this wasn't possible with the framework, so even though both your iOS and macOS targets relied on the framework, for example, you had to duplicate actual implementations for iOSX64, iOSArm64, and macOSX64, which isn't at all scalable. The JetBrains guys came up with a nice solution and even went one step further to automate things.

With the **Commonizer** tool, you can create the aforementioned intermediate source sets, and the tool will be smart enough to infer the common dependencies and create the related abstract/actual declarations for you. It's even smarter than you'd expect because it can also infer the subtle differences between the different POSIX dependencies.

At this point, you may be wondering why we don't use it for any shared code. This is because it works with subtle or no differences at all; it was designed specifically for use cases such as the **POSIX** library, where there are many abstractions to make, but the actual declarations are mostly or completely the same. Also, if it did generate all the shared code, then it would lose its key multiplatform factor; that is, giving developers the flexibility to write shared code while working natively on the platform-specific questions.

If you'd like to fully understand how Commonizer works, I highly recommend this talk by Dmitry Savvinov: https://youtu.be/5QPPZV04-50.

If we'd like to achieve the intermediate source sets shown in *Figure 2.4*, all we have to do is add the following to **Gradle** (if you don't know what Gradle is, you'll find out in the next chapter, but, in brief, it's just a build tool):

```
val appleMain by creating {
    dependsOn(commonMain)
}
val iosMain by getting {
    dependsOn(appleMain)
    dependencies {
        // Add iOS specific dependencies
    }
}
val macOSMain by getting {
    dependsOn(appleMain)
    dependencies {
        // Add macOS specific dependencies
    }
}
```

> **Note**
>
> Since providing an intermediate iOS source set for iOSArm64 and iOSX64 is so common, in that in most cases you probably don't want to write specific code for the different targets, the framework already provides an Intermediate/shared source set configuration. That's why in the preceding code snippet, we don't have to manually create this iOS source set, only the Apple source set.

We could say that the Kotlin/Native backend compiler is currently living its community test phase. It's important to reason why Kotlin/JVM became a success so that we can compare it to the current state of Kotlin/Native. By doing this, we can see what its future could look like, which gives you a better perspective of what you should expect before investing in learning the technology.

I believe that the support of the JetBrains team and the Kotlin community is really good and that because of the soundness of the KMP approach, the framework will scale well. Even though technologies come and go, the multiplatform approach, as a concept, is probably here to stay.

Before we learn how to build KMM apps in more depth, let's cover the Kotlin/JS backend compiler as well so that you have the full picture and know what to expect if you plan on targeting a JS platform as well (be it the web, Node.js, or any other JS target).

The Kotlin/JS compiler

The Kotlin/JS compiler is the final piece of the puzzle for sharing code between different platforms. There are many use cases in which you could leverage this compiler, as follows:

- Sharing code between the backend and the frontend. If your backend is written in Node.js and you'd like to share code between your backends and frontend, Kotlin/JS can be a great tool.

- Sharing code between mobile platforms and the web, which is a great way to keep your frontend in sync.

So, what do you need to know about the Kotlin/JS backend compiler?

How it works

Kotlin/JS currently targets the **ECMAScript 5 (ES5)** JavaScript standard. As we saw with the Kotlin/JVM and Kotlin/Native compilers, Kotlin/JS has a similar process but produces a different type of executable. In short, it takes Kotlin code that it then translates into JavaScript code, given that the underlying code uses dependencies that can run on JavaScript.

It is currently migrating to the IR compiler infrastructure, so instead of directly generating JavaScript files from the Kotlin source code, it first generates an IR, which then gets compiled to JavaScript.

Kotlin currently also has experimental support for generating .d.ts TypeScript declaration files.

If you've been wondering, Kotlin/JS also allows you to use **npm** dependencies in your Kotlin/JS code. It's as simple as doing the following:

```
dependencies {
    implementation(npm("bootstrap", "5.0.1"))
}
```

For those unfamiliar with npm, its purpose is similar to Cocoapods on iOS and Gradle on Android and Java.

As far as concurrency goes in the JS world, because of its single-threaded nature, you won't experience a similar strict concurrency model to Kotlin/Native and the complexity of coroutines may seem obsolete in the Kotlin/JS world.

Coroutines are one of Kotlin's powerful libraries that enable you to write concurrent, asynchronous code. We'll explore them in more depth in *Chapter 3, Introducing Kotlin for Swift Developers.*

Summary

In this chapter, we looked at the three Kotlin compilers that enable code sharing across different platforms: Kotlin/JVM, Kotlin/Native, and Kotlin/JS.

By now, you should have a better understanding of how KMP uses these compilers, how they work, and how they can enable you to share code across different platforms.

In the next chapter, we'll provide a brief introduction to Kotlin to help bring iOS developers up to speed with the Kotlin world. Then, we'll turn to more practical things and dive into creating a KMM project.

3

Introducing Kotlin for Swift Developers

Before we turn things more practical and start developing our **Kotlin Multiplatform Mobile** (**KMM**) apps, I'd like to make sure everyone has the necessary knowledge to follow the steps. This chapter was designed for iOS and Swift developers, especially for those who don't have a comprehensive knowledge of Kotlin and Gradle. Most of the concepts in Swift can be found in Kotlin as well, and in this chapter, we're going to see how Swift's concepts translate to Kotlin. By the end of this chapter, you should be ready for KMM development, by understanding the core Kotlin concepts and other KMM prerequisites. We will be learning about the following topics:

- Introducing Gradle

- Exploring Kotlin's core features

- Understanding Kotlin coroutines

Technical requirements

You can find the code files of his chapter on GitHub at `https://github.com/PacktPublishing/Simplifying-Application-Development-with-Kotlin-Multiplatform-Mobile`.

Introducing Gradle

When writing a KMM application, you'll be using Gradle to build your shared code. For this reason, it's paramount that you at least know the basics, in order to start developing KMM apps.

Gradle is an open source build automation tool and dependency manager. It is similar to CocoaPods on iOS, while covering a broader purpose than pure dependency management, and it is the build tool on which **Kotlin Multiplatform** (**KMP**) is also based.

Gradle provides its own **domain-specific language** (**DSL**) for writing build scripts, and this DSL is available both in Groovy and Kotlin: `build.gradle` is a build script written in Groovy, while `build.gradle.kts` is written in Kotlin.

We will not have an in-depth description of Gradle as it is a huge topic, and without a doubt, many of us as Android developers use it as someone uses a lightbulb: without extensive knowledge on how it works, it still proves to be useful. If you want to gain a more in-depth view of Gradle, I recommend you start with this page: `https://docs.gradle.org/current/userguide/what_is_gradle.html`.

In the following sections, we're going to talk about some of the basic building blocks of Gradle, as well as the features that you'll likely use throughout writing a shared code for a KMM application.

Gradle runs on the **Java Virtual Machine** (**JVM**), and thus you'll need the **Java Development Kit** (**JDK**) in order to use it. But while running on the JVM, it is not limited to building just Java projects—it is suitable for building native ones too.

The great thing about Gradle is that you can extend its current functionality by using tasks and plugins. Tasks usually represent an atomic step of the build process, and plugins are a collection of certain tasks.

For example, KMP provides its own multiplatform Gradle plugin that enables you to use the Multiplatform DSL for specifying the targets you are supporting (such as Android, iOSX64, and so on), and it defines tasks for compiling and building the specific targets. You can read more about the Multiplatform Gradle Plugin DSL here: `https://kotlinlang.org/docs/mpp-dsl-reference.html`.

Another good example is the CocoaPods Gradle plugin (`https://kotlinlang.org/docs/native-cocoapods.html`) with which you can include pods in your shared KMP code, among other capabilities. We'll talk more about this in *Chapter 5, Writing Shared Code.*

Structure of Gradle

For every Gradle module/project, you'll have a `build.gradle` (or `.kts`) file describing the build steps for that specific module. If you have a multi-module app, you'll likely have a top-level build file containing common configurations and options for all modules/sub-projects.

Let's see what a simplified `build.gradle.kts` file could look like in your project, as follows:

```
plugins {
    kotlin("multiplatform")
    id("com.android.library")
    kotlin("plugin.serialization") version
      Versions.KOTLIN_VERSION
    id("org.jetbrains.kotlin.native.cocoapods")
}


kotlin {
    android() // Configures the android target
    ios() // Configures the iOSX64 and iOSArm64 targets


    sourceSets {
        val commonMain by getting {
            dependencies {
                // Common dependencies
                implementation "org.jetbrains.kotlinx:kotlinx-
                coroutines-core:$coroutinesVersion"
            }
        }
        val androidMain by getting {
            dependencies {
                // Android specific dependencies
            }
        }
```

```
        val iosMain by getting {
            dependencies {
                // Native dependencies (for example native
                database drivers)
            }
        }
    }
}
```

Let's go through the different sections in this build configuration.

Plugins

You can see the `plugins {...}` block at the top of the `build.gradle.kts` file. This specifies which plugins you're going to use in the specific project to configure your build; this block only applies specific plugins. In the preceding example, you can see the following plugins:

- `kotlin("multiplatform")` —Official multiplatform plugin for configuring the multiplatform application
- `id("com.android.library")` —Plugin for configuring the Android target
- `kotlin("plugin.serialization")` —Plugin for serialization
- `id("org.jetbrains.kotlin.native.cocoapods")` —CocoaPods plugin

The `kotlin($pluginName)` plugin declaration format is essentially the equivalent of `id("org.jetbrains.kotlin.$pluginName")`.

Kotlin multiplatform configuration

Next, you can observe the `kotlin {...}` configuration block. This is the top-level block for configuring your multiplatform builds. Inside this block, you'll see the `android()` and `ios()` blocks, which essentially specify which targets you are supporting/targeting in your multiplatform project.

You can see the `sourceSets{}` configuration block as well, which is used for configuring predefined source sets or declaring new ones. Source sets contain the Kotlin source files for the specific target and their dependencies.

One thing to note: Gradle needs to know in which repositories to look for to find the specific plugins and dependencies of your project. If you have a single-module project, you can specify this in the same build script. For multi-module projects, this usually goes into a common configuration in a top-level build.gradle.kts file, as illustrated in the following code snippet:

```
// Top-level build file where you can add configuration
    options common to all sub-projects/modules.
buildscript {
    repositories {
        google() // Google Android repository
        mavenCentral() // Maven Central
    }
    dependencies {
      classpath("com.android.tools.build:gradle:agpVersion")
      // Adds the Android Gradle Plugin to the build-script
      classpath
      classpath("org.jetbrains.kotlin:kotlin-gradle-
      plugin:$kotlinVersion") // Adds the Kotlin Gradle
      Plugin to the build-script classpath
    }
}

allprojects {
    repositories {
        google()
        mavenCentral()
    }
}
```

In the preceding top-level Gradle configuration, the following actions occur:

- The `buildscript {...}` block configures the repositories where Gradle should look for certain plugins—in this case, the Android Gradle plugin and the Kotlin Gradle plugin.

- Under the `dependencies {...}` block, you can see that the two previously mentioned plugins get added to the classpath; this is what enables you to later apply specific plugins in your subprojects.

- The `allprojects {...}` block defines repositories across all the subprojects; these are the repositories where Gradle will look to find specific dependencies specified in your subprojects.

Now that you have a basic understanding of Gradle, we should cover some of the core concepts of Kotlin that you'll likely come across when developing KMM or KMP applications.

Exploring Kotlin's core features

Kotlin is an **object-oriented programming** (**OOP**) language with many functional programming features. In this section, we'll go over its main features so that you have a basic understanding of how to express yourself in Kotlin later on. Throughout this chapter, we'll be comparing Kotlin to Swift in terms of these core concepts.

Null safety

Having null references in code proved to be an underestimated factor of error proneness in older languages, such as Java and **Objective-C** (**Obj-C**), thus it has probably earned its *billion-dollar mistake* tag (`https://en.wikipedia.org/wiki/Tony_Hoare#Apologies_and_retractions`).

The purpose and solution are pretty much similar in Kotlin and Swift, with slight syntax and naming differences. They both aim to provide a type system that eliminates the danger of null references. Swift introduced **optional** types, which can be found as **nullables** in Kotlin.

Let's see some code in action, in order to compare unwrapping optionals in Swift to handling nullables in Kotlin. Both Kotlin and Swift use ? to specify a nullable/optional type, as illustrated in the following code snippet:

```swift
// Swift optional
var optional: String? = nil
// Kotlin nullable
var nullable: String? = null
```

Let's see how you would unwrap this optional in Swift, as follows:

```swift
// Swift unwrapping
if let string = optional {
    print(string.count)
} else {
    print("I'm nil")
}
```

Kotlin seems to give you more flexibility on how you handle nullables; we could make it similar to Swift with the following syntax:

```kotlin
// Handling nullables in Kotlin
nullable?.let { string ->
    println(string.length)
} ?: run {
    println("I'm null")
}
```

Kotlin has this nice operator, ? : , also called the Elvis operator, for specifying default values or operations in cases when the value of the nullable is actually null.

Depending on our use case, though, we could make the preceding code much more concise, as follows:

```kotlin
// Handling Nullables in Kotlin
println("${nullable?.length ?: 0 }")
```

Data classes

A data class in Kotlin is a special class that is designed to hold some data. It's similar to Swift's structs, but it doesn't have **copy-on-write** (**COW**) functionality and has the following characteristics:

- At least one constructor parameter, which is also a property of the class
- `equals()`/`hashCode()` is derived by the compiler, based on these constructor parameters
- A derived `copy()` function for cloning a data class and potentially overriding some of its properties

Let's see the following code snippet for a demonstration of this:

```
data class Name(val firstName: String, val lastName:
  String) {
```

Here, we declared a class, with the intent of holding specific information—in this case, a person's name. Going forward, we'll probably need a way to compare two persons' names. Luckily, Kotlin helps out with this, as illustrated here:

```
// Compiler derived functionalities (might be slightly more
complex in reality)
fun equals(other: Name) = firstName == other.firstName &&
  lastName == other.lastName


// Copy with default parameters
fun copy(firstName: String = firstName, lastName: String =
  lastName) = Name(firstName = firstName, lastName =
  lastName)


// Example of using copy
fun createChild(firstName: String) = copy(firstName =
  firstName)
}
```

As you can see from the preceding code snippet, Kotlin not only generates a way for us to compare two names but also generates a `copy()` function, which becomes quite handy when we want to create a copy of a certain class by modifying one part of the contained information only.

In the preceding example, you can see that "creating a child" in this case consists of creating a copy of a given name by keeping the `lastName` and modifying the `firstName` only.

Extensions

Extensions are a great way of adding functionality to classes of which you don't have control. Both Kotlin and Swift support extensions, so let's see what the syntax looks like, as follows:

```swift
// Swift extension example
extension String {
    func appendIf(condition: Bool, suffix: String) ->
  String {
        if(condition){
            return self + suffix
        } else {
            return self
        }
    }
}


// Kotlin extension example
fun String.appendIf(condition: Boolean, suffix: String) =
  if (condition) this + suffix else this
```

Functional programming features

Kotlin, as well as Swift, has great support for functional programming. Both have first-class support for features such as function types, **higher-order functions** (**HOFs**), and lambdas.

We'll not present a detailed overview of all the possibilities; we'll just see a collection-processing example in both languages, as follows:

```swift
// Map signature in Swift
func map<T>(_ transform: (Element) throws -> T) rethrows ->
  [T]
// Example usage of map in Swift
let words = ["Kotlin", "Swift", "are", "both", "beautiful"]
```

```
let letterCounts = words.map { $0.count }
```

```
// Map signature in Kotlin
inline fun <T, R> Iterable<T>.map( transform: (T) -> R):
 List<R>
// Example usage of map in Kotlin
val words = listOf("Kotlin", "Swift", "are", "both",
 "beautiful")
val letterCounts = words.map { it.length }
// or
val letterCounts = words.map { word -> word.length }
```

Objects

Implementing singletons (classes that can have only one instance) is made effortless in Kotlin, whereby everything is basically taken care of by Kotlin. By declaring an `object` instance, you instantly get a singleton, as illustrated here:

```
// Kotlin object
object MySingleton {
    var someState: State
}
```

However, you should probably be aware that in most cases, you don't really want a singleton, or even if you do, you would like to leave the scoping of objects to a **dependency injection** (**DI**) framework instead of the language.

COW in Kotlin

COW is a handy computing technique implemented in Swift that can be a performance enhancer from a macro-optimization standpoint. In Swift, when you copy a struct, Swift essentially delays the copying only when it's needed. For example, this way, if you copy over a large dataset, until you modify the copy, it won't actually have a memory footprint and the behavior will still be the same.

While Kotlin, apart from some syntax differences, looks really similar to Swift, there is no `struct` equivalent in Kotlin and you'll be missing the COW functionality of Swift when writing Kotlin code.

Now that we've covered most of Kotlin's core features, there is one more topic left— one that facilitates asynchronous programming, an essential aspect of mobile development: coroutines in Kotlin.

Understanding Kotlin coroutines

Asynchronous programming is at the heart of mobile development. In order to write efficient applications, leveraging the async capabilities of the framework and language you are using makes all the difference.

Coroutines are my absolute favorite language feature of Kotlin because of their expressiveness and how easy it is to express your asynchronous development needs in a concise way.

In this section, we will be covering the basic concepts of coroutines and compare them to Swift's **async/await** and **Combine** patterns.

Suspend functions

Coroutines are basically suspendable tasks that can suspend and resume execution, and they are not bound to any particular thread.

When you're writing asynchronous code, you generally need to think about the following two things:

- Which task needs asynchronous attention and has to be suspendable
- How you combine asynchronous tasks with the rest of your code

Let's see how you mark any task that needs asynchronous execution first: by writing **suspend functions**.

Whether you're calling a backend **application programming interface** (**API**) request, doing a heavy computation, or performing any other long-running task that could block the **user interface** (**UI**) or main thread, you probably want to mark that execution as suspendable, something that can suspend the execution. Let's see an example here:

```
suspend fun getSuccessRates() =
  remoteSource.getStudents().map { it.calculateSuccessRate()
  }
```

In the preceding example, `remoteSource.getStudents()` returns information about the students (such as name and their grades) from a given backend API by running a network request, after which we calculate the success rate (the probability that the student passes) of each student, supposing that this operation is a costly algorithm that involves complex calculations.

> **Note**
>
> The way the preceding example code calculates each individual's success rate could be improved upon, and we'll get back to this later.

The `suspend` keyword marks the function as suspendable and gives the responsibility to the caller to handle how it would like to run a suspended function. Let's see how this can be done.

In order to understand how you can combine these suspended functions with the rest of the code, we need to tackle `CoroutineScope` first.

CoroutineScope

To better manage the lifecycle and scoping of coroutines, Kotlin introduced `CoroutineScope`; this is what enables structured concurrency in Kotlin. You can create and manage your own `CoroutineScope`, and Android also provides first-party support for most use cases, such as scoping to the lifecycle of `ViewModel` or `View`.

`CoroutineScope` instances can be nested, and canceling a parent scope will cancel all child scopes and all coroutines running under the scopes.

Now that we have a brief understanding of what a `CoroutineScope` is, let's see how you can start to suspend functions and combine them with the rest of your code.

Executing coroutines

In Kotlin, you have mainly two ways for launching Kotlin coroutines: `launch` and `async`.

Let's see some examples, as follows:

```
fun showSuccessRates() {
    coroutineScope.launch {
        val successRates = getSuccessRates()
        // getSuccessRates is a suspend
        function defined previously
```

```
        showUI(successRates) // create UI components based on
        the successRates data
    }
}
```

Generally, we use `launch` when we don't care about the result outside of the coroutine scope, and we can use `async` when we'd like to run operations in parallel and return the result—for example, if we'd like to improve the way we calculate individual student success rates, as illustrated here:

```
suspend fun getSuccessRates() = coroutineScope {
    remoteSource.getStudents().map { async {
        it.calculateSuccessRate() } }.awaitAll()
}
```

In the preceding example, when the code gets to the first item to be mapped, it sees `async`, so it knows that this should be launched separately from the current execution, fires up the success rate calculation, and goes to the next item. When the mapping arrives at the last item, it sees `awaitAll()`, which signals that now is the time where we should wait for all those calculations fired up to be finished before we return it.

There is another aspect of concurrency that we should touch on: multithreading.

Switching threads

Threading is something platform-specific, so whether you're running a multiplatform app on the JVM or Native Kotlin, it will use different platform-specific threading mechanisms. This means also that some of the threading mechanisms may be available to a specific platform only.

Let's see an example of how you could express to the compiler that you'd like to switch threads for a specific operation, as follows:

```
suspend fun getSuccessRates() = withContext(Dispatchers.IO) {
    val students = remoteSource.getStudents()
    withContext(Dispatchers.Default) {
        async { it.calculateSuccessRate() } }.awaitAll()
    }
}
```

In the preceding example, we're switching the execution to the **input/output (I/O)** dispatchers in order to utilize a shared pool of threads designed for offloading networking operations, such as getting the students with a network API call. Second, we're using default dispatchers that were designed for computational work. Since both of these dispatchers work with a shared pool of threads, this means that we might switch threads only once, and Kotlin coroutines know that thread switching is costly and do this only when needed.

> **Note**
>
> `Dispatchers.IO` is currently JVM only. We will see in later chapters how you can offload work to a background thread in the native world.

Unfortunately, there is no official way (yet) for transforming Kotlin suspend functions to async/await, and something like that would need a direct Kotlin <-> Swift **interoperability (interop)**, the development of which is on hold at the time of writing.

Though you can call suspend functions from your Swift code, which will be translated to a completion handler in Swift, we will practically avoid that in most cases, which we'll see in later chapters.

Also, this library might prove to be useful for those looking for a suspend -> async/await mapping later on: `https://github.com/rickclephas/KMP-NativeCoroutines`.

Streams in Kotlin

Streams, and Observable patterns in general, can be a huge component in enabling reactive programming in your apps.

We will have a really quick overview of Kotlin's **Flow** API to give you a sense of what's out there and what the general purpose of it is. For a real deep dive, you should head over to the official documentation at `https://kotlinlang.org/docs/flow.html`.

We'll not cover how you can consume the exposed flows from Kotlin in your Swift/Obj-C code in this chapter, but we'll get back to this in later chapters.

Flow is by default a cold observable, and there are many ways to construct such a cold flow—for example, with the usage of the `flow{}` builder, as illustrated here:

```
suspend fun uploadItem(items: List<Item>) = flow {
    items.forEachIndexed { index, item ->
        remoteSource.upload(item)
```

```
        emit(index + 1)
    }
}
```

In the preceding example, we're basically using a flow that will emit the progress of the upload as we're uploading a number of items. Of course, for an optimal solution, you would probably either use an async/await pattern in Kotlin.

Apart from cold flows, you'll probably need a way to expose state and events in a reactive manner. Let's see how you can do that, as follows:

```
_state = MutableStateFlow<YourState>(DefaultState)
state: StateFlow<YourState> = _state
_events = MutableSharedFlow<YourEvent>()
events: SharedFlow<YourEvent> = _events
// This is how you could update the state
fun updateState(state: YourState) {
    _state.value = state
}
// This is how you can send out an event
fun sendEvent(event: YourEvent) {
    coroutineScope.launch {
        _events.emit(event)
        // Notice that the emit is suspending, this is
        because unless the event is consumed
        // the emitting function will suspend until a
        subscriber comes along
    }
}
// You can collect any Flow like this:
coroutineScope.launch {
    state.collect { state -> }
    // Notice that collect is also a suspendable; this is
    because it will suspend until the flow is
    completed/terminated
}
```

The Flow API is pretty similar to the Combine API in Swift; you can basically think of flows as Publishers.

Summary

In this chapter, we've gone through basic Kotlin and Gradle concepts that are essential in order for you to start sharing code between Android and iOS (and other platforms, potentially) while comparing these concepts to ones that can be found in Swift.

I hope that now you have a proper base knowledge of these tools and you're ready to move forward to more practical things—developing a KMM application.

Section 2 - Code Sharing between Android and iOS

This section offers a practical, hands-on example of how to share code between Android and iOS. After this section, you should have a good understanding of how to write a KMM application.

This section comprises the following chapters:

- *Chapter 4, Introducing the KMM Learning Project*
- *Chapter 5, Writing Shared Code*
- *Chapter 6, Writing the Android Consumer App*
- *Chapter 7, Writing an iOS Consumer App*

4

Introducing the KMM Learning Project

In this chapter, we're going to introduce the KMM project that we're going to implement step by step so that you have a complete understanding of what real-world problems we'll be touching on in the context of the project, what you need before you can start developing, and what implementation decisions were made for which purpose.

We'll cover the following topics in this short chapter:

- Getting to know the project
- Exploring prerequisites
- Understanding the technical decisions

Technical requirements

You can find the code files for this chapter in this book's GitHub repository at `https://github.com/PacktPublishing/Simplifying-Application-Development-with-Kotlin-Multiplatform-Mobile`.

Getting to know our project – Dogify

Our project is called Dogify. In this app, we'll be showing different breeds of dogs with images, with the possibility of favoriting them. We'll get our data from a dog API (`https://dog.ceo/dog-api/`) and then cache it in our local database:

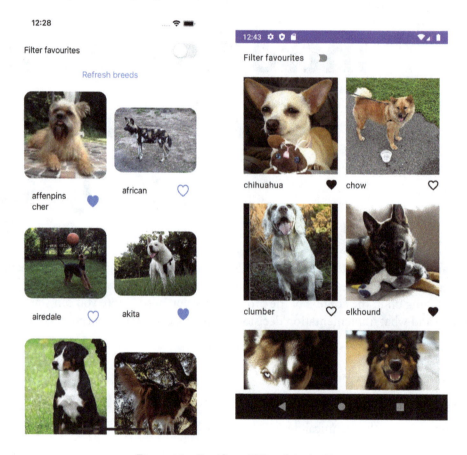

Figure 4.1 – Dogify on iOS and Android

We'd also like to have the possibility of seeing our favorite breeds, which will be stored locally, in our database.

Before we get into the details of our project, let's first understand the objectives that we want our project to meet. Our app/project was designed/selected with the following objectives:

- It is fairly simple to understand and develop.
- It covers most of the common KMM use cases.
- It covers production-related questions.
- It covers the most-used KMM tools.

Since creating a fairly simple project that covers production-related questions as well is quite a challenge, and I'd say conflicting also, we'll leave production questions out of the picture for now.

Rest assured we'll return to it and arm you with tips on where to look for answers to your production-related questions in *Chapter 8, Exploring Tips and Best Practices*, and we'll also tackle some of the most common ones in *Chapter 9, Integrating KMM into Existing Android and iOS Apps*.

We designed our app bearing in mind the things that we would like to see in a KMM app from a technical perspective:

- Setup
- Expect/actual mechanism in work
- Networking
- Database operations
- Multithreading
- Testing

Now, let's go through what you'll need in order to be able to develop and run the project and to understand what we're doing and why.

Exploring prerequisites

First, let's see what the basic skills are that you'll need before starting the following chapters and the work on Dogify.

Skill requirements

The most essential tools and concepts you will need to be familiar with when writing the shared code are the following:

- A basic understanding of Gradle
- An understanding of Kotlin's core concepts

We've covered both of the aforementioned topics in *Chapter 3, Introducing Kotlin for Swift Developers.*

Also, since the shared code will be consumed by an iOS and Android app, knowledge of the following topics, although not necessary, would be good to have:

- **Android Gradle Plugin** (**AGP**) and its **Domain-Specific Language** (**DSL**)
- **Jetpack Compose** (the new declarative UI Toolkit on Android)
- **SwiftUI** (the new declarative UI Toolkit on iOS)
- Swift's **Combine** framework
- Experience with **Xcode**

Since all of the aforementioned is very well documented with multiple sources available and the scope of this book is more about learning to write shared code between mobile apps, we'll not cover any of those exhaustively. However, I'll try to provide a bit of guidance during the development so that even if you don't have extensive knowledge of the topics, you won't feel lost.

Let's jump now to what tools you'll need during the project development.

Required tools

The development of Dogify will consist of three parts on two (or three) different realms:

- Writing the shared code using Kotlin and Gradle
- Writing the Android consumer app using Kotlin and Gradle but with more Android-specific tooling
- Writing the iOS consumer app using Swift

We'll go through the prerequisites for each realm so that you can gain a better perspective of the tooling needs.

Writing the shared code with Kotlin and Gradle

For writing the shared code, you'll first need an **Integrated Development Environment** (**IDE**); in fact, if you're more used to a terminal and you're a **Command-Line Interface** (**CLI**) type of person, you might not "need" it, but it's strongly advised.

You can use **IntelliJ** for starting a KMP app, and in fact, many people do prefer IntelliJ when writing shared code. I personally use it whenever I leave the KMM world and I want to create a project or see a setup that has other targets than just Android and iOS.

You can check out the KMP project wizard in Intellij 2021.1.2, as shown in the following figure:

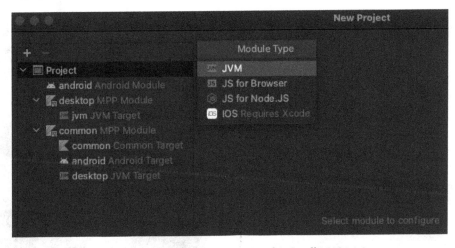

Figure 4.2 – The KMP project wizard in Intellij 2021.1.2

However, for the purposes of this project, I recommend using **Android Studio**, as it will be more suitable for KMM development. You can find the download link for the stable Android Studio at `https://developer.android.com/studio`; I recommend using **Android Studio Arctic Fox**, as it has support for Jetpack Compose: `https://developer.android.com/studio/preview`.

Android Studio is built based on IntellIJ and both have support for Gradle. You won't have to download Gradle manually, since when creating the project, the wizard will add a Gradle Wrapper, which has the purpose of specifying the Gradle version the project will be using (you can change that manually) and fetching it to build your project.

This standardizes the Gradle version that developers use and thus limits inconsistencies when building the project.

You'll also need the **KMM plugin** developed by JetBrains: `https://plugins.jetbrains.com/plugin/14936-kotlin-multiplatform-mobile`. We will be using this plugin mostly to create the new KMM app from scratch, but it can also help with running and debugging the shared code on the iOS target.

> **Important Note**
> The KMM plugin is available only on macOS, as it relies on the Xcode simulators to run and debug the iOS target.

You can install the KMM plugin by going to **Android Studio | Preferences | Plugins**. Then, in the search bar, type in `Kotlin Multiplatform Mobile`:

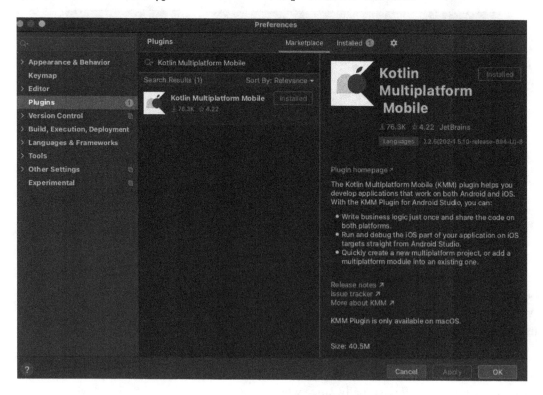

Figure 4.3 – The KMM plugin in IntelliJ/Android Studio

The last step is to make sure you have an installed version of the **Java Development Kit (JDK)** as well. Android Studio has a bundled version of the latest OpenJDK copy, but you can verify the installation by running `java -version` in the terminal. Also, note that Arctic Fox now requires JDK 11+.

The official documentation also has a good description for setting up an environment for KMM: `https://kotlinlang.org/docs/mobile/setup.html`.

Writing the Android consumer app

Now, unless you've started the previous step with IntelliJ, you will have all the tools set up for this stage.

The only thing worth mentioning is that while using the stable Android Studio version is perfectly fine for the shared code, for Jetpack Compose, you will need Android Studio Arctic Fox in order to enjoy the capabilities of Jetpack Compose tooling, such as previews.

At the time of writing, Android Studio Arctic Fox is still in beta, but I do hope that this becomes quickly irrelevant and you can fully leverage Jetpack Compose tooling in stable Studio versions as well.

Writing the iOS consumer app using Swift

In order to be able to run the iOS app, you'll need **Xcode 11+**, as we'll be using SwiftUI to write the UI layer of the app and Swift Package Manager for managing dependencies, which you can download from the App Store.

If you'd like to have one IDE experience, trying out AppCode might also be a good idea (`https://www.jetbrains.com/objc/`); for the purposes of this project, trying out the free trial version will be beneficial.

JetBrains has recently brought out a KMM plugin for AppCode that unifies code completion and code highlighting for both Kotlin and Swift/Objective-C when jumping from one file to another, but debugging is a much more complete package. You can read more about the announcement here: `https://blog.jetbrains.com/kotlin/2021/06/kmm-for-appcode/`.

Now that you know what tools you need, let's check out the structure of the KMM project and the technical choices.

Understanding the technical decisions

In this section, we're going to answer the following questions regarding the project:

- What's our architecture of choice for Dogify?

- Which pieces will we be sharing between the Android and iOS apps?

- What libraries will we be using?

Architecture

Let's address the first question. As we discussed in *Chapter 1*, *The Battle Between Native, Cross-Platform, and Multiplatform*, the purpose of a multiplatform application is to share business, non-UI, or non-platform-specific logic between the different platforms. To make this happen, we need an architecture where the layers facilitate this; that is, where it's easy to divide non-UI layers from the UI layers, essentially.

Luckily, **clean architecture** suits this well, and we'll be implementing one version of it. You can read more about Uncle Bob's Clean Architecture here: `https://blog.cleancoder.com/uncle-bob/2012/08/13/the-clean-architecture.html`.

For Dogify, this is what the architecture will look like:

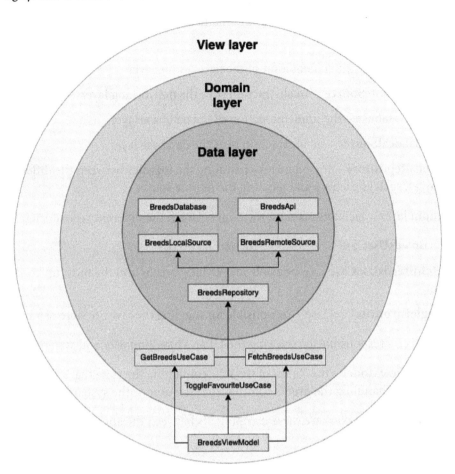

Figure 4.4 – Clean architecture in Dogify

Each layer has different components and different responsibilities:

- **Data layer** – responsible for getting and maintaining the data in the different sources:

  - **BreedsApi** – the implementation of the networking layer

  - **BreedsRemoteSource** – an abstraction over the networking layer

  - **BreedsDatabase** – the implementation of the database layer

  - **BreedsLocalSource** – an abstraction over the database layer

  - **BreedsRepository** – the repository handling the logistics between the different sources, such as caching and fetching the remote source

- **Domain layer** – facilitates communication between the different repositories:

  - **GetBreedsUseCase** – responsible for getting breeds

  - **FetchBreedsUseCase** – responsible for fetching breeds and the most up-to-date data

  - **ToggleFavouriteUseCase** – responsible for toggling the favorite state of a breed

- **View layer** – responsible for showing the UI and handling user and system events:

  - **BreedsViewModel** – the "brain of the UI," containing the state that should be shown and handling the events coming from the user or the system

Now, let's jump to how we'll share these components between the apps.

The shared code

We'll share both the data layer and the domain layer:

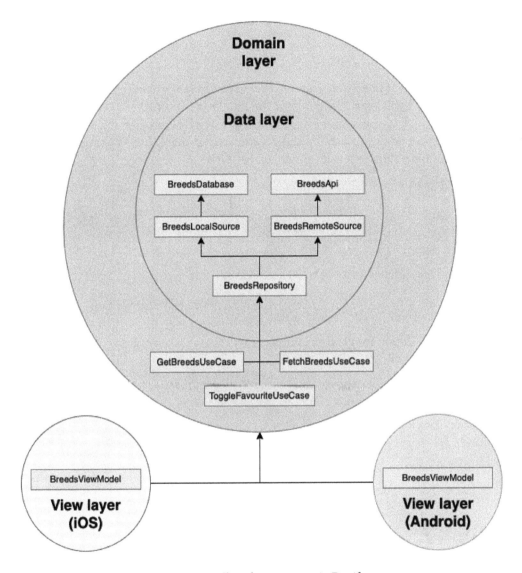

Figure 4.5 – Shared components in Dogify

As you can see in the preceding figure, we will be sharing the data and domain layers across the apps. We will then create a different view layer for iOS and for Android, which communicates or uses this shared layer.

Let's discuss now what libraries we are going to leverage to solve the common use cases.

Library choices

I'll list some of the major library choices so that you can gain a better picture of the tech stack we'll be using. I will explain the specific library choices in *Chapter 5*, *Writing Shared Code*, and also provide alternatives in *Chapter 8*, *Exploring Tips and Best Practices*, while going through an overview of the full palette of currently available KMM libraries (it's probably impossible not to leave out something, but I'll try).

In Dogify, we'll be using the following libraries:

- Kotlin Coroutines, the `native-mt` branch for multithreaded coroutines: `https://github.com/Kotlin/kotlinx.coroutines/tree/native-mt`

- Koin for **Dependency Injection (DI)**: `https://insert-koin.io/`

- Ktor for networking: `https://kotlinlang.org/docs/mobile/use-ktor-for-networking.html`

- Kotlinx serialization for parsing: `https://github.com/Kotlin/kotlinx.serialization`

- SQLDelight for database operations: `https://github.com/cashapp/sqldelight`

- Jetpack Compose for implementing the UI on Android: `https://developer.android.com/jetpack/compose`

- SwiftUI for implementing the UI for iOS: `https://developer.apple.com/xcode/swiftui/`

Summary

By now, you should have a clear understanding of the KMM project we will be building together, its architecture, what will be shared between Android and iOS, as well as which libraries we will use along the way.

Thus, I hope you're ready and eager to start coding Dogify in the next chapters.

5
Writing Shared Code

We're going to jump into implementing Dogify. We'll start from the core of the application – the business logic – in this chapter, which we're going to share across the Android and iOS applications in the following chapters. We're going to focus mainly on the following topics in this chapter:

- Initial project setup
- Fetching data from the Dog API
- Persisting data in a local database

Technical requirements

You can find the code files present in this chapter on GitHub, at `https://github.com/PacktPublishing/Simplifying-Application-Development-with-Kotlin-Multiplatform-Mobile`.

Initial project setup

By now you should have all the tools ready to start developing Dogify. In case you missed something, feel free to explore *Chapter 4, Introducing the KMM Learning Project*, where I described the tools you'll need.

So, let's start creating the project. Open the Android Studio New Project wizard (**Android Studio | New Project**). From the **Phone and Tablet** tab, select **KMM Application**:

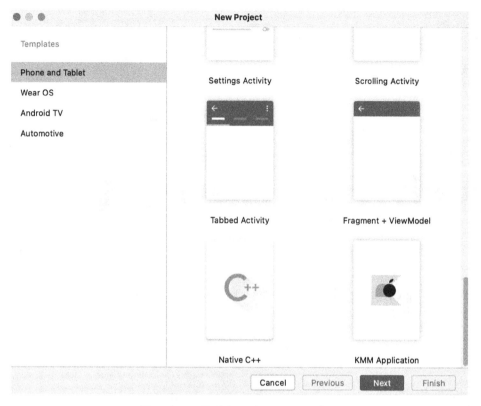

Figure 5.1 – KMM Application template

If you can't see such a template, you're probably missing the KMM Plugin for Android Studio. You can find more about how to install it in *Chapter 4, Introducing the KMM Learning Project.*

Now in the next section, you'll see a configuration page. We'll be using Minimum SDK: API 23 and Kotlin scripts for Gradle build files, so make sure that **Use Kotlin script (.kts) for Gradle build files** is checked. Of course feel free to tweak these attributes, though we're going to use these settings throughout the book.

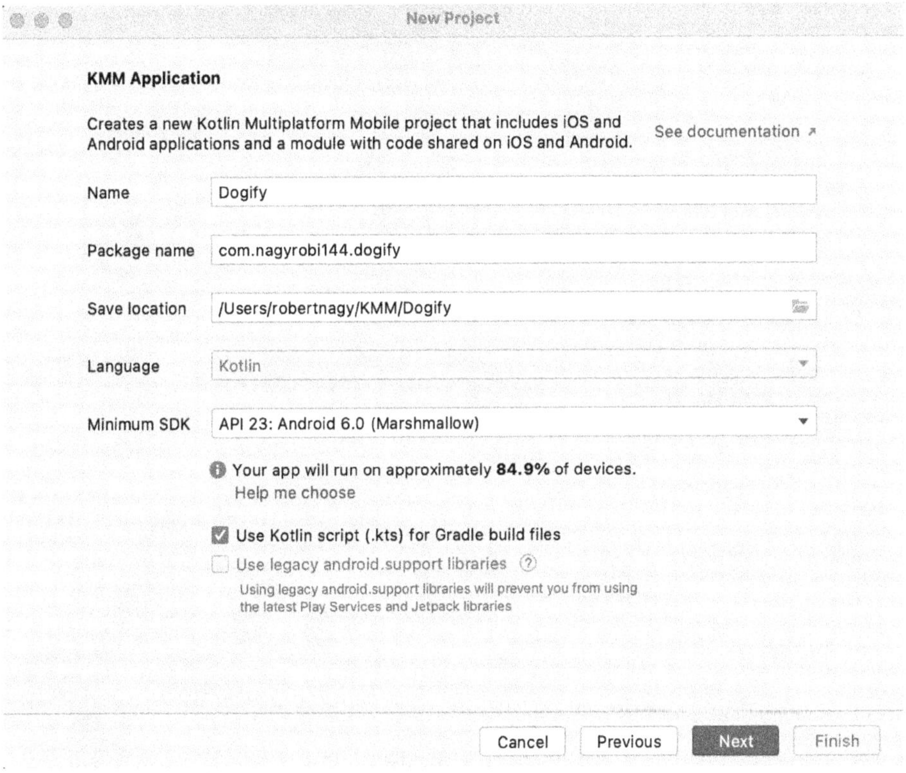

Figure 5.2 – KMM Application first configuration page

After the first configuration page, you'll see yet another configuration page:

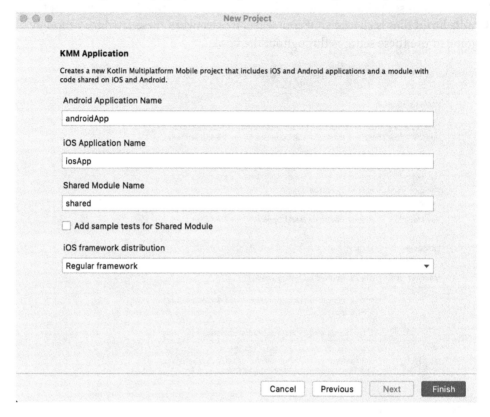

Figure 5.3 – KMM Application second configuration page

On this page you can specify the names for the modules. We're going to use the default names for Dogify. Also, you can choose how you'd like to distribute the Shared code to iOS, with two options currently:

- **CocoaPods dependency manager**
- **Regular framework**

We're going to use the latter, which basically creates a Gradle task called `embedAndSignAppleFrameworkForXcode`, which will be used from your iOS app, as it will consume the shared code in `.framework` format.

Go ahead and click **Finish** and wait for Gradle to set up the project. When Gradle sync has finished, you should be able to run the application:

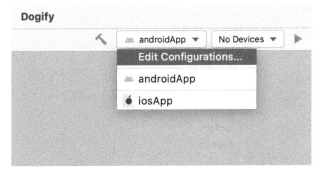

Figure 5.4 – KMM Run configurations for Android and iOS

You can switch between the configurations and try out the project on both Android and iOS.

> **Note**
>
> You need Xcode and a simulator set up in order to be able to run the iOS app from Android Studio. In recent versions the KMM plugin has become much more reliable, though for me what helps when issues occur is to start the simulator from Xcode first, then I am able to consistently run the iOS app from Android Studio as well.

After running you should see **Hello Android {androidVersion}!** and **Hello iOS {iOSVersion}!** on your devices.

Let's examine this initial setup now and do a bit of cleanup before we start.

Project structure

This is how the project structure should look like after creating the KMM application with the KMM project template:

Figure 5.5 – Dogify project structure

As you can see, this is a multi-module Gradle setup, where we have a top-level `build.gradle.kts` for common project configurations, as well module-specific build configurations in the form of individual `build.gradle.kts` files.

The `shared` module contains three `sourceSets`:

- `androidMain`

- `commonMain`

- `iosMain`

You can find the configuration of these source sets in the `build.gradle.kts` of the `shared` module. In short, the meaning of these source sets in a KMP environment is that in the `commonMain` source set, we implement our shared code and potentially define `expected` declarations, while the `androidMain` and `iosMain` source sets will have to contain the `actual` platform-specific implementations. This way KMP will intelligently replace the expected declarations with the actual implementations for the target that it's building for at the given time, as discussed in *Chapter 1, The Battle Between Native, Cross-Platform, and Multiplatform.*

If you open the `build.gradle.kts` of the `shared` module, you should see the following code:

```
plugins {
    kotlin("multiplatform")
    id("com.android.library")
}
```

This is the place where we apply the specific plugins. We need the multiplatform plugin in order to create the multiplatform configurations for the shared code and the Android library plugin so that the shared code can be properly set up as an Android library when consumed from an Android app.

The following section configures the shared code, with two targets, Android and iOS:

```
kotlin {
    android()
    val iosTarget: (String, KotlinNativeTarget.() -> Unit)
    -> KotlinNativeTarget =
        if
        (System.getenv("SDK_NAME")?.startsWith("iphoneos")
        == true)
            ::iosArm64
        else
            ::iosX64

    iosTarget("ios") {
        binaries {
            framework {
                baseName = "shared"
            }
```

```
            }
        }
        sourceSets {
            val commonMain by getting
            val commonTest by getting {
                dependencies {
                    implementation(kotlin("test-common"))
                    implementation(kotlin("test-annotations-
                        common"))
                }
            }
            val androidMain by getting
            val androidTest by getting {
                dependencies {
                    implementation(kotlin("test-junit"))
                    implementation("junit:junit:4.13.2")
                }
            }
            val iosMain by getting
            val iosTest by getting
        }
    }
}
```

This section essentially does two things:

- It specifies the targets that we are going to use in our multiplatform module, an Android and an iOS target respectively.

- It configures the source sets. You can see commonMain, androidMain, and iosMain source sets, and a -Test source set for each. (The reason why you can't see the test folders in the project structure yet is that we didn't tick the checkbox at the start when using the project wizard, which would've generated those as well.)

As you can see, the iosTarget is going to be different, based on if we are using a simulator or a real device to run the shared code: iosArm64 for a real device and iosX64 for a simulator.

Then you can see the `baseName = "shared"` command, which specifies the name of the local framework that will be generated for the iOS app. In case you've run the app with the iOS configuration, you can see that under `shared/build/bin/ios/debugFramework` there is a `shared.framework`.

The last section is as follows:

```
android {
    compileSdkVersion(30)
    sourceSets["main"].manifest.srcFile
    ("src/androidMain/AndroidManifest.xml")
    defaultConfig {
        minSdkVersion(23)
        targetSdkVersion(30)
    }
}
```

This has the Android library-specific configurations.

Let's do a cleanup of the **Hello {platform}** messages and set up the project for Dogify:

1. Delete the `Platform.kt` files in all source sets: `commonMain`, `androidMain`, and `iosMain`.

2. Delete the `Greeting.kt` file in the `commonMain` source set.

There are multiple strategies to choose from when testing your shared code, such as the following:

- Test it only on one target when developing, and when ready test it on all targets.

- Always test it on all targets.

- Hybrid or in-between: run it only on one target, but test platform-specific implementations on all targets.

This is something for you or your team to decide when working on a real project, and it could depend on your project setup and preferences. Testing every change on all targets can be time-consuming but you probably won't end up having to rewrite anything major in your shared code. We will cover more of this topic in *Chapter 9, Integrating KMM into Existing Android and iOS Apps*.

For this exercise, we'll be testing the shared code on Android as we're writing the shared code. Once we've finished it, we'll try it out on iOS and make some changes if necessary.

In *Chapter 4*, *Introducing the KMM Learning Project*, we introduced the architecture and components of Dogify, where the two platform-specific View layers will communicate with the shared Domain layer. More specifically, the Views will interact with a specific use case (otherwise known as Interactors).

Before we start actually implementing, we'll do the following:

1. Create the API contract for the shared code in the form of barebone use cases.
2. Set up dependency injection.

Let's look at these steps in detail.

1. Creating the barebone use cases

Under the commonMain source set, in the model directory, create the following **Data Transfer Object (DTO)**:

```
data class Breed(val name: String, val imageUrl: String,
    val isFavourite: Boolean = false)
```

These are the three main information pieces that we will be transmitting to the UI about a given dog breed:

*   The breed's name
*   A URL to an image of the breed
*   Whether it's a favorite breed or not

Also, in the directory, create the following three Kotlin classes:

*   FetchBreedsUseCase
*   GetBreedsUseCase
*   ToggleFavouriteStateUseCase

These are the three main actions that the apps will run. For now, we will just return some mocked data.

Both getting and fetching the breeds will return us a list of the previously created `Breed`, the difference being that while fetching breeds always returns fresh data from remote, getting breeds returns breeds from the cache if available or performs a fetch:

```
class GetBreedsUseCase {
    suspend fun invoke(): List<Breed> = listOf(Breed("Test
    get", ""), )
}
```

The fetch operation will be different later on in the real implementation, where it'll fetch data from remote instead of getting it from the cache:

```
class FetchBreedsUseCase {
    suspend fun invoke(): List<Breed> = listOf(Breed("Test
    fetch", ""), )
}
```

Toggling the favorite state of a breed will need the breed as a parameter, and it will have no return type:

```
class ToggleFavouriteStateUseCase {
    suspend operator fun invoke(breed: Breed) {
    }
}
```

You might have noticed that we were overriding the `invoke` operator function in our code. For those who aren't familiar with this Kotlin pattern, it just makes invoking a use case a bit nicer – you'll see what I mean after a couple of lines.

2. Setting up the dependency injection

As discussed, we're going to be using Koin – `https://insert-koin.io/`. Let's work through these steps:

1. First, we'll need to specify the dependency under the `commonMain` source set in the `build.gradle.kts`:

    ```
    val commonMain by getting {
        ...
            dependencies {
                api("io.insert-koin:koin-core:3.1.2")
    ```

```
          }
    ...
}
```

We'll declare the dependency as an `api` as a convenience since we also need to do a Koin setup in the consumer Android app as well.

2. Now create a `KoinModule` under the `di` package in the `commonMain` source set, with the following code:

```
private val usecaseModule = module {
    factory { GetBreedsUseCase() }
    factory { FetchBreedsUseCase() }
    factory { ToggleFavouriteStateUseCase() }
}
private val sharedModules = listOf(usecaseModule)
fun initKoin(appDeclaration: KoinAppDeclaration = {})
  = startKoin {
    appDeclaration()
    modules(sharedModules)
}
```

Here we described the "bean definitions", or how we'd like our use cases to be instantiated, where `factory` means a new instance will be created whenever the dependency is injected.

We also expose the `initKoin` method, where the caller can extend on the current Koin configuration.

To be able to test the shared code on Android, go ahead and do the following:

A. Create an application class and initialize Koin:

```
class DogifyApplication: Application() {
    override fun onCreate() {
        super.onCreate()
        initKoin()
    }
}
```

B. Log or print out the results of the shared module API.

I've just simply modified the `greet()` message:

```
suspend fun greet() =
    "${FetchBreedsUseCase().invoke()}\n" +
    "${GetBreedsUseCase().invoke()}\n" +
    "${ToggleFavouriteStateUseCase().invoke(Breed
    ("toggle favourite state test", ""))}\n"

class MainActivity : AppCompatActivity() {
    override fun onCreate(savedInstanceState: Bundle?) {
        super.onCreate(savedInstanceState)
        setContentView(R.layout.activity_main)

        val tv: TextView =
            findViewById(R.id.text_view)

        lifecycleScope.launch {
            tv.text = greet()
        }
    }
}
```

Since the use cases are suspend functions, you'll need a coroutine scope as well to run those. You can see that I've used `lifecycleScope`, which requires the `androidx.lifecycle:lifecycle-runtime-ktx` dependency.

The full code is available on the `05/01-initial-setup` branch.

Now that we have everything set up, let's start implementing our shared code, first by fetching the needed data.

Fetching data from the Dog API

To implement shared networking, we will be using Ktor (`https://kotlinlang.org/docs/mobile/use-ktor-for-networking.html`) and kotlinx.serialization (`https://github.com/Kotlin/kotlinx.serialization`).

First, let's add the dependencies:

1. Add the following dependencies to the common source set:

    - `io.ktor:ktor-client-core:$ktorVersion`

    - `io.ktor:ktor-client-json:$ktorVersion`

    - `io.ktor:ktor-client-logging:$ktorVersion`

    - `io.ktor:ktor-client-serialization:$ktorVersion`

    - `org.jetbrains.kotlinx:kotlinx-serialization-core:1.2.1`

2. We also need to apply the kotlinx-serialization plugin, by adding `kotlin("plugin.serialization") version kotlinVersion` to the plugins block in the shared module's `build.gradle.kts`.

3. And we'll also need to add a dependency for the actual Ktor clients on the different platforms:

    - Add `io.ktor:ktor-client-android:$ktorVersion` to the Android source set.

    - Add `io.ktor:ktor-client-ios:$ktorVersion` to the iOS source set.

Now let's configure our shared Ktor client:

1. We'll start by creating a base class for this configuration. There is probably a better pattern for this, but for the purposes of this example we'll refrain from overengineering, so go ahead and create `KtorApi` under the `commonMain` source set in the `api` directory. We'll make this class as follows:

    ```
    internal abstract class KtorApi {
    ```

    You'll notice that we'll pay attention to using internal modifiers where needed, not only to limit exposing the internals of our shared module, but also because this actually has an impact on the output binary of the Kotlin/Native compiler. We will discuss this in *Chapter 8, Exploring Tips and Best Practices*.

2. Then we'll configure kotlinx-serialization's JSON parsing:

```
private val jsonConfiguration = Json {
        prettyPrint = true
        ignoreUnknownKeys = true
}
```

We want the JSON to be formatted in a human-readable way and the parsing not to fail in case of unknown keys.

3. Then we'll configure our Ktor client:

```
val client = HttpClient {
        install(JsonFeature) {
                serializer =
                    KotlinxSerializer(jsonConfiguration)
        }
        install(Logging) {
                logger = Logger.SIMPLE
                level = LogLevel.ALL
        }
}
```

We'll use `HttpClient`, which under the hood will use the different Android or iOS `HttpClientEngine` from the platform-specific dependencies that we've added.

We also specify the serializer for Ktor to be Kotlinx's Serializer with our previously defined JSON configuration and configure the logging.

4. Now as our last step, we'll expose a method for configuring the API endpoints:

```
fun HttpRequestBuilder.apiUrl(path: String) {
        url {
                takeFrom("https://dog.ceo")
                path("api", path)
        }
}
```

Let's see now how we can get the data with our configured Ktor client:

1. Let's create the DTOs that define how we'd like to parse the data. Create
   `BreedsResponse` and `BreedImageResponse` in the `commonMain` source set
   in the `api/model` directory:

```
@Serializable
internal data class
  BreedsResponse(@SerialName("message") val breeds:
  List<String>)
```

```
@Serializable
internal data class BreedImageResponse(
    @SerialName("message")
    val breedImageUrl: String
)
```

The two DTOs defined here are self-explanatory in terms of how and what we're
parsing out about the responses.

2. Now create a `BreedsApi` under the `commonMain` source set in the `api` directory
   that extends our previously configured `KtorApi`:

```
internal class BreedsApi : KtorApi() {
    suspend fun getBreeds(): BreedsResponse =
    client.get {
        apiUrl("breeds/list")
    }
    suspend fun getRandomBreedImageFor(breed: String):
    BreedImageResponse = client.get {
        apiUrl("breed/$breed/images/random")
    }
}
```

Ktor internally uses the native-mt version of coroutines from `version 1.4+` and it
provides "main-safety" as it switches to a background thread when running the requests.

However, I'd personally consider it a best practice not to rely on a third-party
implementation for this and make sure your `Local-` and `RemoteSources` are
responsible from this perspective.

We're going to introduce an abstraction over the `DispatcherProvider` in coroutines for two reasons:

- Testing.

- Kotlin/Native doesn't have an I/O dispatcher at the moment and we'll need to use the default dispatcher to move tasks to a background thread for iOS. Since we don't want to compromise on Android, we'll want this to be platform-specific and use the classic I/O dispatcher.

Before we create the abstraction, we need to add a dependency on the native multithreaded coroutines in the common source set: `org.jetbrains.kotlinx:kotlinx-coroutines-core:${coroutinesVersion}-native-mt`.

Let's create the following `DispatcherProvider` file under the `util` package in the `commonMain` source set:

```
interface DispatcherProvider {
    val main: CoroutineDispatcher
    val io: CoroutineDispatcher
    val unconfined: CoroutineDispatcher
}
internal expect fun getDispatcherProvider():
  DispatcherProvider
```

As you can see, we've introduced our first expected declarations in our shared code, so we'll need to define the actual implementations for these on both Android and iOS.

Create a `DispatcherProvider` file under the `util` package in the `iosMain` source set:

```
internal actual fun getDispatcherProvider():
  DispatcherProvider = IosDispatcherProvider()
private class IosDispatcherProvider : DispatcherProvider {
    override val main = Dispatchers.Main
    override val io = Dispatchers.Default
    override val unconfined = Dispatchers.Unconfined
}
```

As you can see, we're using the default dispatcher for I/O-related tasks in Kotlin/Native.

Now let's create a `DispatcherProvider` file under the `util` package in the `androidMain` source set:

```
internal actual fun getDispatcherProvider():
  DispatcherProvider = AndroidDispatcherProvider()
private class AndroidDispatcherProvider:
  DispatcherProvider{
    override val main = Dispatchers.Main
    override val io = Dispatchers.IO
    override val unconfined = Dispatchers.Unconfined
}
```

Here, you can see that we'll be using the same I/O dispatcher as we would've probably used for I/O-related tasks on Android.

Now let's create the definition for our `BreedRemoteSource` that will basically have the purpose of creating an abstraction over our third-party networking library. You can add it under the `repository` package in the common source set:

```
internal class BreedsRemoteSource(
    private val api: BreedsApi,
    private val dispatcherProvider: DispatcherProvider
) {
    suspend fun getBreeds() =
      withContext(dispatcherProvider.io) {
        api.getBreeds().breeds
    }

    suspend fun getBreedImage(breed: String) =
      withContext(dispatcherProvider.io) {
        api.getRandomBreedImageFor(breed).breedImageUrl
    }
}
```

As you can see our remote source does two main things above the actual networking API. It makes sure that we're using the I/O dispatcher and it unwraps the data.

We'll also need to combine the two pieces of information, the list of breed names with the image URLs. Let's create the `BreedRepository` under the same `repository` package:

```
class BreedsRepository internal constructor(
    private val remoteSource: BreedsRemoteSource
) {
    suspend fun get() = fetch()
    suspend fun fetch() = supervisorScope {
        remoteSource.getBreeds().map {
            async { Breed(name = it, imageUrl =
            remoteSource.getBreedImage(it)) }
        }.awaitAll()
    }
}
```

Notice the `supervisorScope`, which is used to make sure that a failure of a child job doesn't cancel the execution of the parent.

Since we don't have anything cached, both the get and fetch operations will be identical and we're just going to fetch the list of breeds from the API, then fetch a random image for each of those. In case you're confused how `async` works, we discussed that in *Chapter 3, Introducing Kotlin for Swift Developers*.

As a last step, let's connect these things and test it out in our Android app. Let's add the `BreedsRepository` dependency to both `FetchBreedsUseCase` and `GetBreedsUseCase` – here's how the latter should look after this update:

```
class GetBreedsUseCase: KoinComponent {
    private val breedsRepository: BreedsRepository by
        inject()
    suspend fun invoke(): List<Breed> =
        breedsRepository.get()
}
```

Notice that we're not using Koin's construction injection here, we're using `KoinComponent` instead, and this is for two reasons:

- It's more convenient than injecting into our Android app's view layer.
- It's even more convenient than injecting into our iOS app's view layer with Koin.

The drawback, of course, is that it's harder to test these use cases, as we need to overwrite the bean definitions in order to provide mock dependencies.

As the final step, we need to update our Koin module and extend it with the new component's bean definitions. After the modifications, this is how it will look:

```
private val utilityModule = module {
    factory { getDispatcherProvider() }
}
private val apiModule = module {
    factory { BreedsApi() }
}
private val repositoryModule = module {
    single { BreedsRepository(get()) }
    factory { BreedsRemoteSource(get(), get()) }
}
private val usecaseModule = module {
    factory { GetBreedsUseCase() }
    factory { FetchBreedsUseCase() }
    factory { ToggleFavouriteStateUseCase() }
}
private val sharedModules = listOf(usecaseModule,
    repositoryModule, apiModule, utilityModule)

fun initKoin(appDeclaration: KoinAppDeclaration = {}) =
    startKoin {
        appDeclaration()
        modules(sharedModules)
    }
```

If you run the app on Android you should see the long list of breeds. PS: Don't forget to add the required Android internet permissions. I always forget to do that.

The full code is available on the 05/02-fetching-data branch.

Now that we have some data in our hands, let's see how we can persist it in a local database.

Persisting data in a local database

We'll be using SQLDelight (`https://github.com/cashapp/sqldelight`) for Dogify, though there are other possibilities including NoSQL solutions as well, such as Realm (`https://github.com/realm/realm-kotlin`) and Kodein-DB (`https://github.com/Kodein-Framework/Kodein-DB`).

In this section, we're going to do the following:

1. Explore how to set up SQLDelight in a multiplatform module.
2. Implement the `BreedsLocalSource`.
3. Connect the remaining API functionality to our database.

Let's dive in.

Exploring how to set up SQLDelight in a multiplatform module

As always, the first thing is to add the needed dependencies, so we need to do the following chores:

1. Add the plugin to the `classpath`: add `classpath("com.squareup.sqldelight:gradle-plugin:$sqlDelightVersion")` to the top-level `build.gradle.kts`.
2. Apply the plugin to our shared module, by adding `id("com.squareup.sqldelight")` to the `plugins` block.
3. Add the `"com.squareup.sqldelight:runtime:$sqlDelightVersion"` and `"com.squareup.sqldelight:coroutines-extensions:$sqlDelightVersion"` dependencies to the `commonMain` source set. The second one is needed so that we can make the persisted data the single source of truth as a stream of data, in the form of a Kotlin Flow.
4. We need to add the platform-specific SQL drivers, so add `"com.squareup.sqldelight:android-driver:$sqlDelightVersion"` to the Android source set and the `"com.squareup.sqldelight:native-driver:$sqlDelightVersion` to the iOS source set.

Now, with the use of the SQLDelight Gradle plugin, we can configure our database by adding the following to the Gradle configuration:

```
sqldelight {
    database("DogifyDatabase") {
        packageName = "com.${yourGroupName}.dogify.db"
        sourceFolders = listOf("sqldelight")
    }
}
```

In the previous code snippet, we've specified the name for the database, and we've pointed to the location that should contain the SQL statements for creating and interacting with the database. We shall write those statements now – first let's create `Breeds.sq` under `commonMain/sqldelight.com.${yourGroupName}.dogify.db`.

Now, in order for SQLDelight to be able to create the SQL table and queries that we need, we will add the following:

```
CREATE TABLE breeds(
    name TEXT NOT NULL,
    imageUrl TEXT NOT NULL,
    isFavourite INTEGER AS Boolean DEFAULT 0
);

insert:
 INSERT OR REPLACE INTO breeds(name, imageUrl, isFavourite)
 VALUES (?, ?, ?);

update:
 UPDATE breeds SET imageUrl = ?, isFavourite = ? WHERE name
 = ?;

selectAll:
SELECT * FROM breeds;

clear:
 DELETE FROM breeds;
```

Here, ? signals the parameters of the generated function.

Now if you build the shared module you should be able to see the generated
`DogifyDatabase` and `BreedsQueries` under the shared module's `build/
generated/sqldelight/code/DogifyDatabase` directory.

As you recall, we've added the two platform-specific SQLite driver dependencies to the
Android and iOS source sets. We need to create an instance of this `DogifyDatabase`
that's based on the platform-specific drivers. How should we solve this?

Yes, I hope you guessed it, we're going to create an expected declaration. Also,
since we need an Android `context` we'll try to be clever and leverage Koin's
`androidContext()` functionality. How? First, let's create the expected declaration as
an extension function on Koin's `Scope`:

```
internal expect fun Scope.createDriver(databaseName:
   String): SqlDriver
```

Now, the actual implementation for iOS is pretty simple:

```
internal actual fun Scope.createDriver(databaseName:
   String): SqlDriver =
     NativeSqliteDriver(DogifyDatabase.Schema, databaseName)
```

For Android, let's first add the koin-android dependency `"io.insert-koin:koin-
android:$koinVersion"` to the Android source set (I've used `api` as a convenience
method, so that the consumer Android app module can use this same dependency as well).

Now we can define the actual implementation for creating the SQLite driver on Android
as follows:

```
internal actual fun Scope.createDriver(databaseName:
   String): SqlDriver =
     AndroidSqliteDriver(DogifyDatabase.Schema,
       androidContext(), databaseName)
```

Note that you also need to make sure to define this `androidContext()` in your app:

```
class DogifyApplication : Application() {
    override fun onCreate() {
        super.onCreate()
        initKoin {
            androidContext(this@DogifyApplication)
        }
```

```
        }
    }
```

Now we are able to provide the bean definition for the `DogifyDatabase` in our Koin module:

```
single { DogifyDatabase(createDriver("dogify.db")) }
```

Now that you have the `DogifyDatabase` set up, let's see how we can create the `BreedsLocalSource` on top of it.

Implementing the BreedsLocalSource

Our `BreedsLocalSource` has similar responsibilities to `BreedsRemoteSource`, the main difference being that it communicates with our persistence layer instead of a networking layer.

You can add the following implementation to the `commonMain` source set's `repository` directory:

```
internal class BreedsLocalSource(
    database: DogifyDatabase,
    private val dispatcherProvider: DispatcherProvider
) {

    private val dao = database.breedsQueries

    val breeds = dao.selectAll().asFlow().mapToList()
        .map { breeds -> breeds.map { Breed(it.name,
        it.imageUrl, it.isFavourite ?: false) } }

    suspend fun selectAll() =
      withContext(dispatcherProvider.io) {
          dao.selectAll { name, imageUrl, isFavourite ->
            Breed(name, imageUrl, isFavourite ?: false) }
              .executeAsList()
    }

    suspend fun insert(breed: Breed) =
      withContext(dispatcherProvider.io) {
          dao.insert(breed.name, breed.imageUrl,
```

```
        breed.isFavourite)
    }
    suspend fun update(breed: Breed) =
      withContext(dispatcherProvider.io) {
          dao.update(breed.imageUrl, breed.isFavourite,
          breed.name)
    }
    suspend fun clear() =
      withContext(dispatcherProvider.io) {
          dao.clear()
    }
}
```

You probably noticed the following line:

```
    val breeds = dao.selectAll().asFlow().mapToList()
        .map { breeds -> breeds.map { Breed(it.name,
          it.imageUrl, it.isFavourite ?: false) } }
```

This maps the underlying data in the database to a Kotlin Flow, so that our data becomes a reactive stream.

Connecting our database to the rest of the components

First, we'll update our repository so that it caches the fetched breeds properly, and retrieves and updates this cache when needed. Let's extend our BreedsRepository:

```
class BreedsRepository internal constructor(
    private val remoteSource: BreedsRemoteSource,
    private val localSource: BreedsLocalSource
) {
    val breeds = localSource.breeds

    internal suspend fun get() =
      with(localSource.selectAll()) {
          if (isNullOrEmpty()) {
              return@with fetch()
          } else {
```

```
                    this
          }
      }
    internal suspend fun fetch() = supervisorScope {
        remoteSource.getBreeds().map {
            async { Breed(name = it, imageUrl =
              remoteSource.getBreedImage(it)) }
        }.awaitAll().also {
            localSource.clear()
            it.map { async { localSource.insert(it) }
        }.awaitAll()
        }
    }

    suspend fun update(breed: Breed) =
        localSource.update(breed)
}
```

In the preceding code, you can see the following three actions updated:

- `get()` now first checks if some cached breeds exist in the local source, and if yes, then it returns them, otherwise it starts a fetch.

- `fetch()` gets the data from the remote source, then it saves the returned breeds.

- `update(breed)` simply updates the given breed.

Now that we have the update function implemented, we can also implement `ToggleFavouriteStateUseCase`:

```
class ToggleFavouriteStateUseCase: KoinComponent {
    private val breedsRepository: BreedsRepository by
      inject()

    suspend operator fun invoke(breed: Breed){
        breedsRepository.update(breed.copy(isFavourite =
          !breed.isFavourite))
    }
}
```

As you can see, we're basically switching the current favorite state of the given breed.

The last thing we need to do is to update the repository module in Koin:

```
private val repositoryModule = module {
    single { BreedsRepository(get(), get()) }
    factory { BreedsRemoteSource(get(), get()) }
    factory { BreedsLocalSource(get(), get()) }
}
```

Now you should be able to test the code for yourself by running the Android app. The full code is available on the 05/03-persisting-data branch.

> **Note**
> This variation of a "clean architecture" isn't meant to impose any architectural preference or boilerplate on anyone. I personally always feel that it's easier to read well-structured code, and this is an effort to define such a code, which has worked out for me and the people I've worked with so far.

Summary

Having worked through this chapter, you've had the benefit of some hands-on experience of how to set up a KMM project and have seen how you can leverage KMP's expect/actual mechanism. We also covered how you can write a shared networking layer between Android and iOS as well as how to write a shared persistence layer between Android and iOS.

In the next chapter, we're going to put this shared layer to work on Android and see how it's able to perform.

6
Writing the Android Consumer App

Now that we've implemented the shared code, we should put it to the test. We'll start with the easier step first; that is, consuming the shared module from the Android code. This chapter will be a more concise one as teaching Android development is outside the scope of this book. With that said, I consider it important to see how that shared KMM code can be consumed by the targeted platforms.

In this chapter, we'll cover the following topics:

- Setting up the Android module
- Tying the Android app to the shared code
- Implementing the UI on Android

Technical requirements

You can find the code files for this chapter in this book's GitHub repository at
`https://github.com/PacktPublishing/Simplifying-Application-Development-with-Kotlin-Multiplatform-Mobile`.

Setting up the Android module

Since we tested part of the shared code in *Chapter 5*, *Writing Shared Code*, we have already done most of the setup. Let's go through what we need to set up before implementing the Android app.

Enabling Jetpack Compose

We'll be using Android's new UI Toolkit: Jetpack Compose. So, first, we'll need to enable it. You can find the official setup guide here: `https://developer.android.com/jetpack/compose/setup#add-compose`.

To enable Jetpack Compose, we'll need to add the following configurations to the `build.gradle.kts` file of the `androidApp` module, under the `android{}` configuration block:

1. Enable the `compose` build feature:

    ```
    buildFeatures {
            compose = true
    }
    ```

2. Make sure both the Kotlin and Java compilers target Java 8:

    ```
    compileOptions {
            sourceCompatibility = JavaVersion.VERSION_1_8
            targetCompatibility = JavaVersion.VERSION_1_8
    }
    kotlinOptions {
            jvmTarget = "1.8"
    }
    ```

3. We also need to specify the version of the Kotlin compiler extension to be used:

    ```
    composeOptions {
            kotlinCompilerExtensionVersion =
            composeVersion
    }
    ```

The last step is to add all the dependencies we'll need to implement Dogify on Android.

Adding the necessary dependencies

Make sure that you add the following dependencies to the `build.gradle.kts` file of the `androidApp` module:

1. Add the following dependency of our shared module:

```
implementation(project(":shared"))
```

2. Now, we must add `AppCompat` and various Kotlin extensions. Note that instead of relying on `AppCompat`, you could just use ComponentActivity; see `https://twitter.com/joreilly/status/1364982668371329025?s=20`:

```
implementation("androidx.appcompat:appcompat:1.3.0")
    // Android Lifecycle
    val lifecycleVersion = "2.3.1"
implementation("androidx.lifecycle:lifecycle-
    viewmodel- ktx:$lifecycleVersion")
implementation("androidx.lifecycle:lifecycle-
    runtime-ktx:2.4.0-alpha02")
    // Android Kotlin extensions
implementation("androidx.core:core-ktx:1.6.0")
kotlinOptions {
        jvmTarget = "1.8"
}
```

3. Now, let's add the Jetpack Compose UI, Foundation, Activity, and the various tooling support that's required, such as Previews:

```
implementation("androidx.activity:activity-
    compose:1.3.0-rc02")
implementation("androidx.compose.ui:ui:$composeVersion
    ")
    // Tooling support (Previews, etc.)
implementation("androidx.compose.ui:ui-
        tooling:1.0.0-rc02")
    // Foundation (Border, Background, Box, Image,
    Scroll, shapes, animations, etc.)
    implementation("androidx.compose.
foundation:foundation
    :$composeVersion")
```

4. Now, add Jetpack Compose Material Design and the necessary icons:

```
// Material Design
implementation("androidx.compose.material:material:$co
mposeVersion")
// Material design icons
implementation("androidx.compose.material:material-
    icons-core:$composeVersion")
implementation("androidx.compose.material:material-
    icons-extended:$composeVersion")
```

5. Finally, add the Accompanist coil for image loading and swipe to refresh the capabilities:

```
val accompanistVersion = "0.13.0"
implementation("com.google.accompanist:accompanist-
    coil:$accompanistVersion")
    implementation("com.google.accompanist:accompanist-
    swiperefresh:$accompanistVersion")
```

The full code is available at **06/01-android-module-setup**.

> **Note**
>
> There is a common pattern for sharing dependency versions between the shared and Android modules. In most cases, this is done by storing it in a Kotlin file/object in buildSrc. I've refrained from this pattern for the following reasons:
>
> • I wanted to keep the practical chapters as simple and to the point as possible.
>
> • I'm not sure how I feel about the pattern and about tying the dependency versions for these modules together. It can be great in example projects, but production KMM apps are not that likely to live in the same repository. This is typically the case when the project is adopting KMM in already existing apps.

Now that we have all the dependencies in place, we are ready to consume the shared code.

Tying the Android app to the shared code

We'll be using a simple `ViewModel` pattern to interact with the shared code and expose the needed data and actions to our UI, based on Android's architecture `ViewModel` to leverage some life cycle functionality provided by the framework.

We'll create a simple `MainViewModel` class in the `androidApp` module. Let's go through the implementation step by step.

First, let's think about what dependencies this `ViewModel` has:

```
class MainViewModel(
    breedsRepository: BreedsRepository,
    private val getBreeds: GetBreedsUseCase,
    private val fetchBreeds: FetchBreedsUseCase,
    private val onToggleFavouriteState:
      ToggleFavouriteStateUseCase
) : ViewModel() {
```

Since we'll be communicating with the shared code, we'll make use of the three use cases for running the specific actions, and we'll listen to the stream containing the breeds from `BreedsRepository`.

Now, let's look at what will we expose from the `ViewModel` layer to `View`. We will do this with the help of Kotlin's Flows, exposing the whole state of the UI in multiple pieces of information:

1. The "state" will tell the UI some information about the UI, if we have any data, if something went wrong, or whether we're currently loading the data:

    ```
    private val _state = MutableStateFlow(State.LOADING)
    val state: StateFlow<State> = _state


    private val _isRefreshing = MutableStateFlow(false)
    val isRefreshing: StateFlow<Boolean> = _isRefreshing


    enum class State {
          LOADING,
          NORMAL,
          ERROR,
          EMPTY
    }
    ```

2. We must send out events when, for example, a certain action has failed (such as marking a breed as a favorite):

```
private val _events = MutableSharedFlow<Event>()
val events: SharedFlow<Event> = _events
enum class Event {
    Error
}
```

3. We must also provide information about whether we are currently filtering out favorite breeds only:

```
private val _shouldFilterFavourites =
    MutableStateFlow(false)
val shouldFilterFavourites: StateFlow<Boolean> =
    _shouldFilterFavourites
```

4. Finally, we need to provide the list of breeds, which depends on the breeds coming from `BreedsRepository` and whether the user wants to filter out favorite breeds only:

```
val breeds =
        breedsRepository.breeds.combine
        (shouldFilterFavourites) { breeds,
        shouldFilterFavourites ->
            if (shouldFilterFavourites) {
                breeds.filter { it.isFavourite }
            } else {
                breeds
            }.also {
                _state.value = if (it.isEmpty())
                State.EMPTY else State.NORMAL
            }
        }.stateIn(
            viewModelScope,
            SharingStarted.WhileSubscribed(),
            emptyList()
    )
```

Now, let's see what actions we can expose for the UI in the form of functions:

1. The first thing we need is a refresh action so that users can trigger a force refresh of the underlying data:

```
fun refresh() {
        loadData(true)
}
```

2. We also need a way for users to switch between whether they'd like to see their favorite breeds only or not:

```
fun onToggleFavouriteFilter() {
    _shouldFilterFavourites.value =
        !shouldFilterFavourites.value
}
```

3. Next, we need an action for marking or unmarking a favorite breed:

```
fun onFavouriteTapped(breed: Breed) {
        viewModelScope.launch {
            try {
                onToggleFavouriteState(breed)
            } catch (e: Exception) {
                _events.emit(Event.Error)
            }
        }
}
```

4. We also need a trigger for getting the data when `ViewModel` is initialized:

```
init {
        loadData()
}
```

5. At this point, our `loadData()` function should look like this:

```
private fun loadData(isForceRefresh: Boolean = false) {
        val getData: suspend () -> List<Breed> =
            { if (isForceRefresh) fetchBreeds.invoke()
            else getBreeds.invoke() }
```

```
        if (isForceRefresh) {
            _isRefreshing.value = true
        } else {
            _state.value = State.LOADING
        }

        viewModelScope.launch {
            _state.value = try {
                getData()

                State.NORMAL
            } catch (e: Exception) {

                State.ERROR
            }
            _isRefreshing.value = false
        }
    }
```

Essentially, we're just checking if we should do a force refresh or not and calling the appropriate use case from the shared code. Based on the result of this operation, we update the "state."

This completes the implementation of our `MainViewModel`. The last thing we need to do is make sure that Koin knows how to inject this `ViewModel`. For this, we'll create an `AppModule` file that contains the following bean definition of our `ViewModel`:

```
val viewModelModule = module {
    viewModel { MainViewModel(get(), get(), get(), get()) }
}
```

We also need to make sure Koin knows about this newly declared module by adding it to our initialization:

```
initKoin {
            androidContext(this@DogifyApplication)
            modules(viewModelModule)
}
```

The full code is available at **06/02-consuming-shared-code-android**.

Now that we've tied together our Android app to the shared code, let's try it out by building the UI and testing it.

Implementing the UI on Android

Before we start, I'd like to emphasize that I had conflicting thoughts when I was writing this chapter (as a matter of fact, the whole example project). I wanted to polish the UI as much as possible, try out the new Android 12 splash screen API, make it edge-to-edge, and so on. But at the same time, I didn't want to introduce things without explicitly talking about them in this book as well, and to do that felt out of scope.

So, consider this as me finding an excuse for why the UI looks so barebone.

Now, let's throw some Jetpack Compose code together and see how consuming the shared code can be presented on an Android UI:

1. Let's create a `MainScreen` that will contain our small number of composable components. We'll start by creating the `MainScreen` composable:

    ```
    @Composable
    fun MainScreen(viewModel: MainViewModel) {
        val state by viewModel.state.collectAsState()
        val breeds by viewModel.breeds.collectAsState()
        val events by
          viewModel.events.collectAsState(Unit)
        val isRefreshing by
          viewModel.isRefreshing.collectAsState()
        val shouldFilterFavourites by
          viewModel.shouldFilterFavourites.collectAsState()
    ```

    As you can see, first, we consume all the state-related information that our previously defined `ViewModel` exposes.

2. We'll also need two other states to be maintained by this composable:

    ```
    val scaffoldState = rememberScaffoldState()
    val snackbarCoroutineScope = rememberCoroutineScope()
    ```

3. Now, our root component will be a `Scaffold` with a `SwipeRefresh` so that users can trigger the refresh action with a pull-to-refresh action:

```
Scaffold(scaffoldState = scaffoldState) {
    SwipeRefresh(
        state =
        rememberSwipeRefreshState(isRefreshing =
        isRefreshing),
        onRefresh = viewModel::refresh
    )
```

4. Next, we'll split the screen into two parts – a switch for toggling the favorite breeds filter and one for the content. The latter will either contain the list of breeds, a loading indicator, or an empty/error placeholder:

```
Column(
        Modifier
        .fillMaxSize()
        .padding(8.dp)
        ) {
    Row(
        Modifier
        .wrapContentWidth(Alignment.End)
        .padding(8.dp)) {
            Text(text = "Filter favourites")
    Switch(
        checked = shouldFilterFavourites,
        modifier = Modifier.padding
        (horizontal = 8.dp),
        onCheckedChange = {
        viewModel.onToggleFavouriteFilter() }
                )
            }
    when (state) {
        MainViewModel.State.LOADING -> {
            Spacer(Modifier.weight(1f))
            CircularProgressIndicator(Modifier.align
            (Alignment.CenterHorizontally))
```

```
                    Spacer(Modifier.weight(1f))
                }
    MainViewModel.State.NORMAL -> Breeds(
        breeds = breeds,
        onFavouriteTapped =
            viewModel::onFavouriteTapped
    )


    MainViewModel.State.ERROR -> {
        Spacer(Modifier.weight(1f))
        Text(
            text = "Oops something went wrong...",
            modifier =
            Modifier.align(Alignment.CenterHorizontally)
        )
        Spacer(Modifier.weight(1f))
}
    MainViewModel.State.EMPTY -> {
        Spacer(Modifier.weight(1f))
        Text(
            text = "Oops looks like there are no
        ${if (shouldFilterFavourites) "favourites"
            else "dogs"}",
            modifier =
            Modifier.align(Alignment.CenterHorizontally)
            )
        Spacer(Modifier.weight(1f))
    }
}
if (events == MainViewModel.Event.Error) {
    snackbarCoroutineScope.launch {
        scaffoldState.snackbarHostState.apply {
            currentSnackbarData?.dismiss()
            showSnackbar("Oops something went wrong...")
                }
            }
```

```
            }
        }
    }
```

5. The last component will be the `Breeds` composable, which will show the list of breeds, their images and their names, and an action for marking that breed as a favorite in a grid:

```
@Composable
fun Breeds(breeds: List<Breed>, onFavouriteTapped:
 (Breed) -> Unit = {}) {
    LazyVerticalGrid(cells = GridCells.Fixed(2)) {
        items(breeds) {
            Column(Modifier.padding(8.dp)) {
                Image(
                    painter = rememberCoilPainter(request =
                      it.imageUrl),
                    contentDescription = "${it.name}-image",
                        modifier = Modifier
                            .aspectRatio(1f)
                            .fillMaxWidth()
                    .align(Alignment.CenterHorizontally),
                        contentScale = ContentScale.Crop
                )
        Row(Modifier.padding(vertical = 8.dp)) {
            Text(
                text = it.name,
                modifier = Modifier
            .align(Alignment.CenterVertically)
                )
        Spacer(Modifier.weight(1f))
            Icon(
                if (it.isFavourite)
```

```
                    Icons.Filled.Favorite else
                    Icons.Outlined.FavoriteBorder,
                contentDescription = "Mark as favourite",
                modifier = Modifier.clickable {
                            onFavouriteTapped(it)
                    }
                )
            }
        }
    }
}
```

6. Finally, we must show these components. We'll need to update our `MainActivity` to the following:

```
class MainActivity : AppCompatActivity() {

    private val viewModel by viewModel<MainViewModel>()
    override fun onCreate(savedInstanceState: Bundle?) {
        super.onCreate(savedInstanceState)
        setContent {
            MaterialTheme {
                MainScreen(viewModel)
            }
        }
    }
}
```

Here, we're injecting the `MainViewModel` component we created with Koin's `ViewModel` functionality and setting the `MainScreen` composable as the content of `MainActivity`. If you run this code, you should be able to see the following screen:

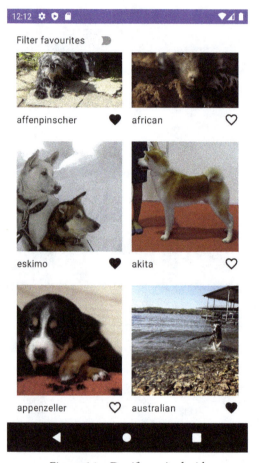

Figure 6.1 – Dogify on Android

The full code is available on branch `06/ui-android`.

Summary

In this chapter, we connected our shared code to our Android application and created the Dogify UI in Jetpack Compose. We also observed that consuming the shared code was as easy as consuming a regular Android library.

In the next chapter, we'll try to do the same for iOS and see if we need to make any modifications to our shared code to make it work and easier to consume via Swift.

7

Writing an iOS Consumer App

After trying out the shared code on Android, the next step will be doing the same for iOS. Spoiler alert—it won't be exactly as seamless as on Android, and we'll need to make some adaptations to the shared code. For this reason, I believe this chapter should offer a good insight into what extra work is actually needed for making **Kotlin Multiplatform** (**KMP**) shared code work on iOS and how seamless it can be with the current tech stack on Swift. This chapter will follow a similar structure to *Chapter 6*, *Writing the Android Consumer App*, and in it, we will explore the following topics:

- Setting up the iOS app

- Tying the iOS app together with the shared code

- Implementing the UI on iOS

Technical requirements

You can find the code files present in this chapter on GitHub, at `https://github.com/PacktPublishing/Simplifying-Application-Development-with-Kotlin-Multiplatform-Mobile`.

Setting up the iOS app

The heavy lifting is done for us by the **Kotlin Multiplatform Mobile (KMM)** plugin we discussed in *Chapter 4, Introducing the KMM Learning Project*.

If you open the iosApp.xcodeproj file in the iosApp module with Xcode and open the **Build Phases** tab for the iosApp target, under the **Run Script** phase, you should be able to see the following command:

```
./gradlew :shared:embedAndSignAppleFrameworkForXcode
```

This is an integrational task that, as the name suggests, embeds and signs a generated Apple framework from the shared code. This task is visible only from Xcode and can't be used from the **command-line interface (CLI)**. The task is illustrated in the following screenshot and you can read more about it at https://blog.jetbrains.com/kotlin/2021/07/multiplatform-gradle-plugin-improved-for-connecting-kmm-modules/:

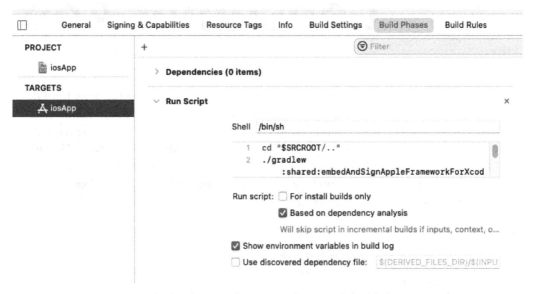

Figure 7.1 – embedAndSignAppleFrameworkForXcode build phase in Xcode

After you delete the Greeting from the ContentView, you should be able to build the project with the shared.framework included.

Also, if you import the shared module, you should be able to see the public API that we defined for the shared code—for example, FetchBreedsUseCase, as illustrated in the following screenshot:

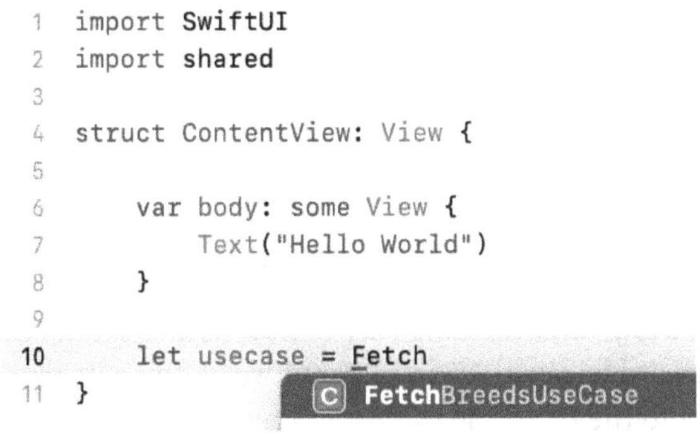

```
1    import SwiftUI
2    import shared
3
4    struct ContentView: View {
5
6        var body: some View {
7            Text("Hello World")
8        }
9
10        let usecase = Fetch
11    }                    C  FetchBreedsUseCase
```

Figure 7.2 – Autocomplete for shared code framework

Note that in production apps, you might not go down on the same pathway for distributing the shared code to iOS; we'll talk about this more in *Chapter 8*, *Exploring Tips and Best Practices*, and *Chapter 9*, *Integrating KMM into Existing Android and iOS Apps*.

Now that we have the shared code near at hand, we are able to put it to the test and create a Dogify app for iOS on top of it.

Tying the iOS app together with the shared code

If you try to subscribe to the breeds stream from the BreedsRepository repository, you'll see that you can't really initialize this repository from the iOS code. This is because we've made a mistake—we don't really want to deal with Koin from Swift, so we could just migrate to a similar injection pattern that we've used for the use cases, as illustrated in the following code snippet:

```
class BreedsRepository: KoinComponent {
    private val remoteSource: BreedsRemoteSource by
    inject()
    private val localSource: BreedsLocalSource by inject()
```

Now, if you try to call some of our use cases, you can see that you can call suspend functions, as illustrated in the following screenshot, and you'll get back a `CompletionHandler`, since Kotlin 1.4:

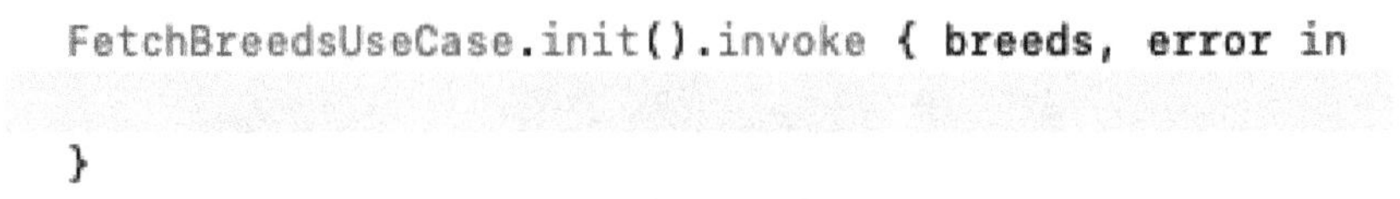

```
FetchBreedsUseCase.init().invoke { breeds, error in

}
```

Figure 7.3 – Calling suspend functions from Swift

Now, there are a couple of concerns about using this approach, as outlined here:

- Collecting Kotlin Flows and running suspend functions from Android is pretty straightforward because Android provides handy extensions for tying **CoroutineScopes** to the **life cycle** of the components, a good example being `viewModelScope`, which we've used to launch suspend functions and collect Flows. Unfortunately, there is no way to specify a **CoroutineScope** when running suspend functions from Swift.

- When launching a coroutine in Kotlin, it returns a `Job`, which you can cancel manually. Since this is also not available in Swift, you can't really cancel a running job from your Swift code.

I believe these are the cases when the power of KMP surfaces versus cross-platform solutions: you're not driven into a corner as you can just go back to the Kotlin world, handle Kotlin specific things there, and expose an API that can be called from Swift.

You can get creative with the way you handle this, but generally, you would perform the following steps:

1. Create a Native version of your `suspend` function, as follows:

```
suspend operator fun invoke(): List<Breed> =
    breedsRepository.fetch()
val nativeScope = YourScope()
// You could keep a Job reference also in this case
fun invokeNative(
        onSuccess: (breeds: List<Breed>) -> Unit,
        onError: (error: Throwable) -> Unit
    ) {
        try {
            nativeScope.launch {
```

```
                    onSuccess(invoke())
            }
        } catch (e: Throwable) {
            onError(e)
        }
    }
```

2. Expose to Swift a way to cancel the running operation, like this:

```
fun onCancel() {
        nativeScope.cancel()
}
```

This example is just a simplified solution, and many questions could arise, depending on your architecture and preference, such as the following: *Do you handle scopes? If so, how do you handle them? Do you leave it to the shared code, or do you handle individual jobs instead?* We'll talk more about this topic in *Chapter 8, Exploring Tips and Best Practices*.

One thing is pretty clear: with the current solution for consuming suspend functions and Kotlin flows, you will likely end up writing a lot of boilerplate code. This is where code-generation tools come into the picture. Luckily, there are libraries that can help us out with this, such as the following ones:

*   `https://github.com/FutureMind/koru`
*   `https://github.com/rickclephas/KMP-NativeCoroutines`

We'll be using the latter for Dogify.

To set up the `rickclephas/KMP-NativeCoroutines` library, you need to perform the following steps:

1. Apply the `id("com.rickclephas.kmp.nativecoroutines")` version `"0.4.2"` plugin in your shared module's `build.gradle.kts` file.
2. Install CocoaPods by doing the following:

    I.   Open up a terminal in the `iosApp` module directory.
    II. Install CocoaPods if you don't have it already by running the following command: `sudo gem install cocoapods`.
    III. Run `pod init` to initialize the Podfile, then `pod install`.
    IV. Reopen the project, but instead of `iosApp.xcodeproj`, use `iosApp.xcworkspace`.

V. (Optional) We also need to update the KMM plugin run configuration similarly, as otherwise, it'll not find our installed pods. To do this, go to **Edit Configurations…** (on the `iosApp` dropdown next to **Play**), as illustrated in the following screenshot:

iosApp
Pixel 2 API 28
Edit Configurations...
androidApp
iosApp

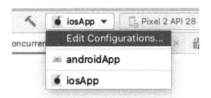

Figure 7.4 – Edit Configurations… location

VI. Update `iosApp.xcodeproj` to `iosApp.xcworkspace` similar to *step IV*. You also need to select the Xcode project scheme to be `iosApp`, as illustrated in the following screenshot (you'll probably need to wait a couple of seconds until it loads):

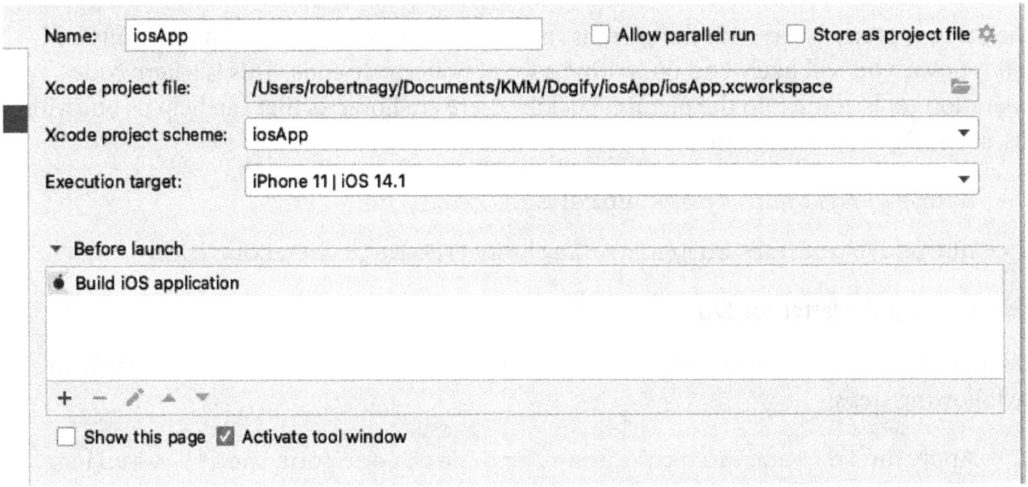

Name: iosApp
Allow parallel run
Store as project file
Xcode project file: /Users/robertnagy/Documents/KMM/Dogify/iosApp/iosApp.xcworkspace
Xcode project scheme: iosApp
Execution target: iPhone 11 | iOS 14.1
Before launch
Build iOS application
Show this page
Activate tool window

Figure 7.5 – Selecting .xcworkspace instead of .xcodeproj

3. Add pod `'KMPNativeCoroutinesRxSwift'` to the Podfile, then run `pod install`.

You'll probably run into an issue, as Xcode won't find the `KMPNativeCoroutinesCombine` module when trying to import it. If you go to **Build Settings | Framework Search Paths**, you'll see that there is only one value: `$(SRCROOT)/../shared/build/xcodeframeworks/$(CONFIGURATION)/$(SDK_NAME)`, which seems to be the result of the KMM plugin's new project wizard. It's not clear why and how the plugin's configuration breaks the CocoaPods settings, but adding the pod's path explicitly using `"${PODS_CONFIGURATION_BUILD_DIR}"` with a recursive search should fix the problem.

Now that we have the library set up, we can focus on tying the shared code to our iOS app. We'll also experience the two most common threading issues, as outlined next, and see how we can fix them:

1. Accessing non-shared/mutable state from another thread

2. Mutating state that is frozen/immutable

First, let's mirror the `MainViewModel` from Android, but now, we're going to use the `Combine` pattern to publish the state.

1. Create a `MainViewModel` Swift class that extends `ObservableObject` and try to consume the shared module, with the help of the `KMPNativeCoroutinesRxSwift` library.

    So, for the use cases that expose `suspend` functions, we'll be using the `createSingle()` helper, while for the Kotlin Flow stream of Breeds, we will be using the `createObservable()` method. Here's an example to illustrate this:

```swift
createObservable(for:
  BreedsRepository.init().breedsNative)
  .subscribe(onNext : { value in
         print("Received value: \(value)")
      }, onError: { error in
         print("Received error: \(error)")
      }, onCompleted: {
         print("Observable completed")
      }, onDisposed: {
         print("Observable disposed")
      })
```

```
createSingle(for:
  getBreeds.invokeNative()).subscribe(onSuccess: {
  value in
          print("Received value: \(value)")
      }, onFailure: { error in
          print("Received error: \(error)")
      }, onDisposed: {
          print("Single disposed")
})
```

2. We also need to initialize Koin, which we'll do in the `init()` method of `ContentView`, as follows:

```
struct ContentView: View {
    private let viewModel: MainViewModel
    init() {
        KoinModuleKt.doInitKoin()
        viewModel = MainViewModel.init()
    }
```

If you run the project now, you'll get the following error:

```
illegal attempt to access non-shared
  org.koin.core.context.GlobalContext
  .KoinInstanceHolder@28557e8 from other thread
```

If you check our use cases and the `BreedsRepository` repository, you'll see that we're using by `inject()` to inject our dependencies. This is a lazy injection method that will only get triggered when we're actually running the Native `GET` request from another coroutine scope. To fix this, we have essentially two options, as follows:

1. Make the Koin `GlobalContext` shareable between threads, which requires freezing, and then requires making sure we're not mutating it.

2. Don't use lazy injection.

Since the second option is much simpler and it's not a big drawback, we'll be using that approach instead. So, we'll need to replace `by inject()` with `= get()` in all of our use cases and in the `BreedsRepository` repository as well.

> **Note**
>
> `KMP-NativeCoroutines` by default uses a coroutine scope with `Dispatchers.Default` when creating `native()` suspend functions. If you want other functionality, you can override it and specify your own coroutine scope. We'll not do this in order to keep things simple.

If you run the app again, you will get another exception again with a long stack trace that starts with this:

```
illegal attempt to access non-shared
   org.koin.core.context.GlobalContext
  .KoinInstanceHolder@28557e8 from other thread
```

If you scroll down, you'll see the root cause, as follows:

```
Caused by:
  kotlin.native.concurrent.InvalidMutabilityException:
  mutation attempt of frozen
  com.nagyrobi144.dogify.api.BreedsApi@2901a8
```

And you can see that the mutation happens for our `HttpClient` in `KtorApi`. Essentially, the problem is similar to the previous one—we're initializing the `HttpClient` in one thread and capturing it in another thread.

In this case, to fix this issue, we'll make the `HttpClient` both a singleton and shareable between threads, as follows:

1. We'll move the `HttpClient` construction from `KtorApi` and make it a `@SharedImmutable`, as follows:

```
private val jsonConfiguration get() = Json {
    prettyPrint = true
    ignoreUnknownKeys = true
}
@SharedImmutable
private val httpClient = HttpClient {
    install(JsonFeature) {
```

```
            serializer =
                KotlinxSerializer(jsonConfiguration)
        }
    install(Logging) {
        logger = Logger.SIMPLE
        level = LogLevel.ALL
        }
    }
```

2. We'll reference this `httpClient` from our `KtorApi`, as follows:

```
        val client = httpClient
```

You can probably do this in a much nicer way, but the main thing is to make the `HttpClient` a singleton and shareable across threads. We've written the code in such a way that it requires as few code changes as possible so that it's easier to follow.

Now, if you run the app again, I'm sorry, but you'll hit another exception, as we can see here:

```
failed with exception:
    kotlin.native.concurrent.InvalidMutabilityException:
    mutation attempt of frozen kotlin.collections.HashMap
```

You can see that the HTTP request has run, and we've even got a response, so the issue probably occurred during parsing. After the due diligence of checking if anyone else has a similar issue with `kotlinx-serialization`, we can find a workaround—we need to use it in our **JavaScript Object Notation (JSON)** configuration, as follows:

```
    useAlternativeNames = false
```

Note that the preceding code is fixed in `kotlinx-serialization` v1.2.2. Until the current concurrency model is updated, you'll likely see similar issues—in some cases, in your own code, and in other cases, in third parties. We'll talk more about how to handle similar issues in *Chapter 8, Exploring Tips and Best Practices*.

Now that we've adapted our shared code to iOS, we're all set to consume the shared code and create a proper UI for it.

To implement the `ViewModel`, proceed as follows:

1. Add the dependencies on the shared module, as follows:

```
private let repository = BreedsRepository.init()
private let getBreeds = GetBreedsUseCase.init()
private let fetchBreeds = FetchBreedsUseCase.init()
private let onToggleFavouriteState =
  ToggleFavouriteStateUseCase.init()
```

2. Then, define the state we'll publish to the UI, as follows:

```
@Published
private(set) var state = State.LOADING
@Published
var shouldFilterFavourites = false
@Published
private(set) var filteredBreeds: [Breed] = []
@Published
private var breeds: [Breed] = []
```

3. Now, let's create actions that we'll be exposing to the UI. We'll start with a `getData` action for getting the data, as follows:

```
func getData(){
    state = State.LOADING
    createSingle(for:
        getBreeds.invokeNative())
        .subscribe(onSuccess: { _ in
        DispatchQueue.main.async {
            self.state = State.NORMAL
        }
    }, onFailure: { error in
        DispatchQueue.main.async {
            self.state = State.ERROR
        }
    }).disposed(by: disposeBag)
}
```

4. Then, we'll create a similar `fetchData` function for refreshing the data, as follows:

```
func fetchData() {
    state = State.LOADING


    createSingle(for:
        fetchBreeds.invokeNative())
        .subscribe(onSuccess: { _ in
            DispatchQueue.main.async {
                self.state = State.NORMAL
            }
        }, onFailure: { error in
            DispatchQueue.main.async {
                self.state = State.ERROR
            }
        }).disposed(by: disposeBag)
}
```

5. Next, we'll create an action for marking or unmarking a favorite breed, as follows:

```
func onFavouriteTapped(breed: Breed) {
    createSingle(for:
        onToggleFavouriteState.invokeNative(breed:
        breed)).subscribe(onFailure: { error in
        // We're going to just ignore
        }).disposed(by: disposeBag)
    }
        } catch (e: Exception) {
            _events.emit(Event.Error)
        }
    }
}
```

As you've seen in the previous code snippets, we're using the following code to make sure we aren't updating the state from a background thread:

```
DispatchQueue.main.async {
}
```

Also, we're using `DisposeBag` to dispose of the running RxSwift `Singles`.

6. We also need to subscribe to the `breeds` stream of the `BreedsRepository` repository, as follows:

```
init() {
    createObservable(for:
        repository.breedsNative)
        .subscribe(onNext: { breeds in
            DispatchQueue.main.async {
                self.breeds = breeds
            }
        }).disposed(by: disposeBag)
```

7. Then, we tie the list of breeds to the flag, representing whether or not the users want to filter their favorite breeds, like so:

```
$breeds.combineLatest($shouldFilterFavourites, {
    breeds, shouldFilterFavourites -> [Breed] in
        var result: [Breed] = []
        if (shouldFilterFavourites){
            result.append(contentsOf:
            breeds.filter{ $0.isFavourite })
        } else {
            result.append(contentsOf: breeds)
        }
        if (result.isEmpty){
            self.state = State.EMPTY
        } else {
            self.state = State.NORMAL
        }
        return result
}).assign(to: &$filteredBreeds)
```

To make sure we have data when `MainViewModel` is initialized, we'll also start getting the data by simply calling the `getData()` function.

Now that our `MainViewModel` is in place, we can prepare the UI as the last step.

Implementing the UI on iOS

We're going to leverage SwiftUI to build the UI for our iOS app, and we'll basically mirror the declarative UI in Jetpack Compose from Android.

We'll also use Kingfisher to load the images, so let's start with the following steps:

1. Update the Podfile by adding pod 'Kingfisher' then running pod install.

2. Next, update the ContentView, as follows:

```
@ObservedObject private var viewModel: MainViewModel

init() {
    KoinModuleKt.doInitKoin()
    viewModel = MainViewModel.init()
}

var body: some View {
    VStack{
        Toggle("Filter favourites", isOn:
            $viewModel.shouldFilterFavourites)
            .padding(16)
        Button("Refresh breeds", action: {
            viewModel.fetchData()} )
            .frame(alignment: .center)
            .padding(.bottom, 16)
        ZStack{
            switch viewModel.state {
            case MainViewModel.State.LOADING:
                ProgressView()
                    .frame(alignment:.center)
            case MainViewModel.State.NORMAL:
                BreedsGridUIView(breeds:
                    viewModel.filteredBreeds,
                    onFavouriteTapped:
                    viewModel.onFavouriteTapped)
            case MainViewModel.State.EMPTY:
                Text("Ooops looks like there are
```

```
                              no breeds")
                       .frame(alignment: .center)
                       .font(.headline)
              case MainViewModel.State.ERROR:
                  Text("Ooops something went
                   wrong...")
                       .frame(alignment: .center)
                       .font(.headline)
                  }
              }
          }
      }
```

We're basically handling the state coming from our `MainViewModel` and showing the appropriate UI elements based on the state.

3. Next, we'll need to implement the `BreedsGridUIView`, which basically sets up the Grid only. This is achieved with the following code:

```
struct BreedsGridUIView: View {
    var breeds: Array<Breed>
    var onFavouriteTapped: (Breed) -> Void = { _ in }

    var body: some View {
        let columns = [
            GridItem(.flexible(minimum: 128, maximum:
                256), spacing: 16),
            GridItem(.flexible(minimum: 128, maximum:
                256), spacing: 16)
        ]
        ScrollView{
            LazyVGrid(columns: columns, spacing: 16){
                ForEach(breeds, id: \.name){ breed in
                    BreedUIView(breed: breed,
                    onFavouriteTapped: onFavouriteTapped)
                }
            }.padding(.horizontal, 16)
```

```
        }
    }
}
```

4. The last step is to show the actual breed in the grid, with an action for marking it as
    a favorite. Here's the code you'll need:

```
struct BreedUIView: View {

    var breed: Breed
    var onFavouriteTapped: (Breed) -> Void = {_ in }

    var body: some View {
        VStack{
            KFImage(URL(string: breed.imageUrl))
                .resizable()
                .scaledToFit()
                .cornerRadius(16)
            HStack{
                Text(breed.name)
                    .padding(16)
                Spacer()
                Button(action: { onFavouriteTapped(breed)
                }, label: {
                    if(breed.isFavourite){
                        Image(systemName: "heart.fill")
                            .resizable()
                            .aspectRatio(1, contentMode: .fit)
                            .frame(width: 24)
                    } else {
                        Image(systemName: "heart")
                            .resizable()
                            .aspectRatio(1, contentMode: .fit)
                            .frame(width: 24)
                    }
                }).padding(16)
```

```
                    }
              }
          }
      }
```

And that's it—we have created a UI for iOS on top of the shared code that we've written in Kotlin.

Summary

In this chapter, we've fixed four issues related to the Kotlin/Native concurrency model. Next, we adapted the Kotlin code to the Native world and the iOS app. We also consumed the shared code in iOS and created a UI for the iOS app in SwiftUI.

The main purpose of this chapter was to provide a glimpse into how you can consume shared code implemented in KMP and how to approach the challenges of making your code work in the Native world.

In the following chapter, we will look at how you can write tests for your shared code.

Section 3 - Supercharging Yourself for the Next Steps

This section will explore tips and best practices about the technology and how to incorporate it into existing production apps.

This section comprises the following chapters:

- *Chapter 8, Exploring Tips and Best Practices*
- *Chapter 9, Integrating KMM into Existing Android and iOS Apps*
- *Chapter 10, Summary and Your Next Steps*

8
Exploring Tips and Best Practices

While we've laid down the basics of Kotlin multiplatform development and also created a minimal KMM app, there is still a lot of ground to be covered. Efficient mobile development is a much more complex game and different aspects also need to be discussed. Uber's almost catastrophic Swift rewrite may be a testament to why choosing a technology is far from being a trivial question: `https://twitter.com/StanTwinB/status/1336890442768547845?s=20`.

Due to this, in this and the upcoming chapters, I will cover the different topics that are affected by a mobile development process based on KMP. More explicitly, in this chapter, we will cover the following topics:

- Testing shared code
- Architectural decisions
- Managing concurrency
- App size best practices

Technical requirements

You can find the code files for this chapter on GitHub at `https://github.com/PacktPublishing/Simplifying-Application-Development-with-Kotlin-Multiplatform-Mobile`.

Testing shared code

Testing shared code in KMP is similar to writing code in KMP: in your shared code, you will have to write platform-agnostic code. This means no third-party testing frameworks or libraries that target a specific platform, JVM, JS, or iOS, such as XCTest or JUnit. Thankfully, KMP already provides a library that targets JVM, JS, and Native: `https://kotlinlang.org/api/latest/kotlin.test/`.

`kotlin.test` provides an `Asserter` abstraction with a `DefaultAsserter` that is dependency-free, but it also provides `JUnitAsserter`, `JUnit5Asserter`, and `TestNGAsserter` so that you can choose the one you'd like to use in your JVM or Android targets.

You can also implement your own `Asserter` implementations for the different platforms if you wish. The same expect/actual mechanism can be used in your tests as well.

But what about which test double (`https://en.wikipedia.org/wiki/Test_double`) you should choose to test your shared code – Stubs, Mocks, Spies, Fakes, or Dummies? This is probably a topic that could span a whole book. If you are someone like me, who likes relying on a third party to create the mocks and spies you use, you'll probably really miss this currently in your shared code.

Unfortunately, there is no mocking library that supports Kotlin/Native. MockK probably has the most potential in supporting it in Kotlin/Native and they have an open issue for it: `https://github.com/mockk/mockk/issues/58`. Honestly, it's not the end of the world; fakes have many benefits over mocks. If you are convinced about MockK and using a mocking library, I highly suggest reading this article: `https://medium.com/@june.pravin/mocking-is-not-practical-use-fakes-e30cc6eaaf4e`.

The last topic to discuss regarding testing is async code and threading. The solution here is simple and many use this pattern in the Kotlin world already:

- Make an abstraction over the dispatchers and have it as a dependency in your components, similar to what we did in `DispatcherProvider` in Dogify.

- Provide a `TestDispatcherProvider` in your tests.

- Use `runBlocking{}` in your tests to run suspend functions.

If you'd like to check out an example of testing, I've included one for Dogify in the `08/01-testing` branch.

As we move forward, we'll see how KMP may affect your architecture.

In the next section, we'll cover how a shared layer can influence the architecture of your apps.

Architectural decisions

There is a wide range of opinions regarding which architecture is the most suitable for traditional apps; there is no one size fits all decision. As we've already experienced, KMP gives you a lot of flexibility in how you plan on organizing your shared code.

Certain things will be influenced by KMP and probably move you in a certain direction (such as choosing fakes over mocks). Before we dive into these constraints, I want to emphasize one key benefit of shared code and KMP. In *Chapter 1, The Battle Between Native, Cross-Platform, and Multiplatform*, we discussed the costs of keeping Android and iOS in sync. While there are best practices to limit working in silos, enforcing a shared architecture is one of the things that teams find to be one of the immediately evident benefits of KMP as it becomes easier to communicate between teams.

Thus, I'd say that an ideal architecture in KMP would be "*one that accomplishes sharing the most of the architectural layers, but still allows room for platforms to shine.*" – Captain Obvious.

Now, the good thing is that if you feel that the architecture that currently fits on iOS may not make sense on Android, you don't have to enforce that approach with KMP. I'd evaluate this decision again before moving forward.

Now, let's discuss what constraints you may face in KMP that may shape your architecture.

Interacting with coroutines

As we saw in Dogify, you can interact with Kotlin coroutines via Swift code. Some teams decide on another path, handling coroutines completely in your shared code.

In this case, I have found that sharing the ViewModel layer, for example, can work well. You can manage the coroutine scopes per ViewModel to expose convenient callbacks for consuming state changes for the UI, as well as a function for clearing the coroutine's scope, when necessary. You can find an example of this at `https://github.com/halcyonmobile/MultiplatformPlayground/blob/master/common/src/commonMain/kotlin/com/halcyonmobile/multiplatformplayground/viewmodel/HomeViewModel.kt`.

While things can get more platform-specific from your ViewModel layer, you can generally make abstractions over these (using resources such as strings, images, and so on). I believe that most apps want better consistency between Android and iOS, so sharing ViewModel layers can be a beneficial step in aligning the platforms.

Sharing resources (such as colors, localization, and so on) may also be something that you'd like to do. For example, if you plan on sharing the ViewModel layer, you'll find that this decision will push you toward sharing resources as well since, in many cases, ViewModel logic exposes the right resources to the view. How easily you can share resources may also influence this decision. Finally, there are libraries out there for this already, such as `https://github.com/icerockdev/moko-resources`.

However, sharing ViewModels may not be suitable in certain situations since you may want a different UI or navigation on the different platforms, so you may find that sharing the ViewModel layer doesn't make sense anymore.

Again, the way I'd think of choosing an architectural pattern and the percentage of shared code would be going for as much alignment and shared code as possible, while also keeping in mind the freedom of the platforms and platform teams. In my opinion, aligning architectures and platforms is beneficial and if it's enforced through shared code, then it will be even better. But if your ways of working with big platform teams are different, you may be biting off more than you can chew.

Interacting with data transfer objects (DTOs)

Kotlin/Native interoperability with Obj-C and Swift is pretty good, but it's not as perfect as writing code in that language. We've already talked about trying to work with Kotlin code in Kotlin when possible. This is, I believe, another pro of the ViewModel layer since most of the data processing is done in your shared Kotlin code; you just expose the UI models.

You may have observed a general pattern here, which is the result of the interoperability capabilities of KMP. This can be generalized to any Kotlin code that is consumed on a specific platform.

Interacting with shared code

In *Chapter 2, Exploring the Three Compilers of Kotlin Multiplatform*, we discussed the interoperability qualities of KMP. While K/N interoperability is pretty good, it probably won't be 100% perfect any time soon. Since KMP gives you a lot of flexibility in terms of how you interact with this shared code, developers can use their creativity and programming skills to leverage the framework to make this interaction more robust. Let's explore what options you have in those cases when you are not satisfied with how you can consume your shared code in Swift/Obj-C.

In general, you have two choices to improve on the experience of working with shared code:

- Move the consumer into the shared Kotlin code too. Consuming Kotlin code in Kotlin is easier.

- Decide not to share that part. Consuming code on the platform/language where it was written is always easier.

We've also discussed KMP's expect/actual declarations. While they sound like a neat feature of multiplatform development, in reality, you may not use them much since they rob you of some flexibility. Why is that, and what may be a better solution?

With expect/actual, you get nice IDE warnings, but the drawback is that you always have to write the platform implementation in Kotlin. But what if you want that implementation to use pure Swift code?

You can still make abstractions by using interfaces (which will be mapped to protocols in Obj-C/Swift) and then provide the implementations based on the running platform by using a DI solution. It may take a bit more effort to tie things together and be able to provide your implementations from Swift. However, once you've done this, it offers a great deal of flexibility regarding how you can write your platform-specific code.

Using interfaces also makes testing easier and increases flexibility as your abstractions won't be tied to platforms – they become more dynamic.

Now, let's look at managing concurrency in KMP apps.

Managing concurrency

> **Important Note**
>
> Starting with Kotlin 1.6.10, the new memory model is enabled by default, with the official multithreaded coroutines library available for Kotlin/Native. This change makes the following overview and the freezing concept in Kotlin/Native obsolete. While you may bump into the freezing model until the new memory model becomes stable, a pragmatic approach would be revisiting/reading up on freezing-related concepts when the need arises.

In the previous chapters, we saw how Kotlin/Native's concurrency model differs from JVM and that while a new model is being made, it will probably take some time until it's stable. In this section, we'll explore some of the more common concurrency issues that people have and what best practices you can follow to avoid them.

As a quick reminder about Kotlin/Native's concurrency rules, you can only share immutable states between threads. This is done at runtime and is referred to as a frozen state. Freezing an object is a one-way operation and can't be reversed; so, once something is frozen, it can't be mutated afterward.

If you need to work around these rules, options are available, such as atomics and thread-isolated states, which are documented at `https://kotlinlang.org/docs/mobile/concurrent-mutability.html`.

In many cases, concurrency "confusion" arises when Kotlin/Native crosses paths with Swift code, especially when using lambdas. As we've discussed previously, for a state to be shared across threads in Kotlin/Native, it needs to be frozen. When you're calling your shared code from Swift, you can make sure that your parameters are frozen by doing the following:

- Freezing them in your shared code.

- Exposing a `freeze()` method to Swift and freezing it before you call the method.

It can also be helpful to use `ensureNeverFrozen()` to make sure that those objects that you wouldn't like to be frozen, will never be frozen. This will cause an exception to occur whenever an attempt is made to freeze that specific object.

Also, keep in mind that freezing can recursively freeze everything that the state touches. In some cases, it may not be that obvious. Here are a couple of short examples to illustrate some common issues that may not be straightforward to debug:

- Switching threads with `withContext()` will capture everything under the function block and freeze it. In this example, `breedName` will be frozen:

```
val breedName = "Vizsla"
withContext(Dispatchers.Default) {
    remoteSource.getImageUrl(breedName)
}
```

- The returned value from another thread will also be frozen. Here, the returned `Breed` will also be frozen:

```
fun getBreed() = withContext(Dispatchers.Default) {
    Breed("Vizsla")
}
```

- Here, the parent class will also be captured, so not only the `breedName` value but its parent class, `GetBreedsUseCase`, will be frozen too:

```
class GetBreedUseCase(val breedName: String){

    suspend operator fun invoke() =
withContext(Dispatchers.Default){

        getBreed(breedName)

    }

}
```

The official documentation has matured recently and now covers some common scenarios as well: `https://kotlinlang.org/docs/mobile/concurrency-overview.html`.

There may be other cases that you may be confused by. The best way to approach these is to do the following:

1. Reread the official documentation.

2. Look at what happens in your code.

3. Ask for help on Stackoverflow or in Slack channels if you can't seem to figure it out. The community is great and grows every day.

Fortunately, the KMP team has started working on a new memory manager that will make these tips deprecated. Let's examine what we know so far about this new memory manager.

New memory manager

Starting with Kotlin 1.6, you can try out the new Kotlin memory manager development preview, which is a simple but not optimal version of the new memory manager. For production apps, we will still have to wait a bit, but with the new memory manager, you can forget about all the restrictiveness of the Kotlin/Native concurrency model and the whole freezing mechanism.

So far, working with this restrictive Kotlin/Native concurrency model has arguably been the hardest part of KMP. This is in part not only because you had to understand the freezing mechanism, but because you also had to construct the shared code with Kotlin/Native's concurrency model in mind, a code smell in shared code. Thus, the concurrency model was a big blocker for adoption, despite Kevin Galligan's best efforts and good materials on educating people on the old memory manager.

I think that when the new memory manager reaches production readiness, it will likely push KMM adoption hugely in the community.

If you've been thinking about how this change will affect your current code, don't worry – unless you're working with some internals, you will probably be safe and you'll be advised to remove code that freezes your objects. Unless you explicitly require certain objects to be frozen in your code.

To learn more about the changes that have been made in the Kotlin memory manager, take a look at the following resources:

- *Kotlin/Native Memory Management Roadmap*: `https://blog.jetbrains.com/kotlin/2020/07/kotlin-native-memory-management-roadmap/`

- *Kotlin/Native Memory Management Update*: `https://blog.jetbrains.com/kotlin/2021/05/kotlin-native-memory-management-update/`

- *Try the New Kotlin/Native Memory Manager Development Preview*: `https://blog.jetbrains.com/kotlin/2021/08/try-the-new-kotlin-native-memory-manager-development-preview/`

Apart from developer convenience, there is another important aspect that can be crucial to the success of your app. No matter how nice a weather app you write, it probably won't get many downloads if it's 1 GB, for example. Now, let's take a quick look at KMP's impact on app size.

App size best practices

Many people are interested in how big of an impact shared KMP code can have on their iOS app's size. This will depend on what kind of logic it contains and what third parties your KMP code uses.

Generally speaking, since Kotlin and Swift are both fairly static languages, there is not much difference between the binary size of a Kotlin class and a Swift class. So, what's the catch?

As you may recall from *Chapter 2*, *Exploring the Three Compilers of Kotlin Multiplatform*, we discussed how your Obj-C/Swift code can see this native output of the Kotlin/Native compiler by generating Obj-C adapters. Since Obj-C is a more dynamic language, it negatively impacts binary size.

So, what can you do to manage binary size? The simple answer is to limit how many Obj-C adapters you use. How? Mark everything that you don't need to expose through your shared code as internal.

If you notice that your binary size has increased in a non-linear, exponential fashion, then it would be a good idea to revisit your visibility modifiers and ensure you only expose what you need.

Summary

In this chapter, we talked about how to test shared KMP code, how your architectural decisions may be influenced by shared code, and covered various concurrency and app size best practices.

In the next chapter, we'll focus on other production and adoption-related questions such as team and repository structure, tooling, and adoption tips.

9
Integrating KMM into Existing Android and iOS Apps

So far, we've discussed the KMM tech and how it works. While adopting KMM into apps in production, you'll probably have some questions about the implications of KMM on the logistics of app development. In this chapter, we're going to focus on the DevOps perspective of KMM and discuss the following topics:

- Deciding on a mono repository or a shared library
- Exploring team structure and tooling
- Learning some adoption tips

By the end of this chapter, you should be prepared to take the next steps and adopt KMP in your projects.

Deciding on a mono repository or a shared library

In this section, we're going to reason about the pros and cons of the following two repository setups:

- **Mono Repository**: This is where the shared code is just a module/submodule. This is the choice of most KMM/KMP example projects.

- **Multiple Repositories**: This is where the shared code is like a *library* that's consumed by the different platforms; that is, Android and iOS. Most production apps will likely see this option as more attractive.

Let's start by looking at mono repositories.

Mono repository

In a mono repository structure, your shared code, the Android app, and the iOS app are all contained in the same repository, as shown in the following diagram:

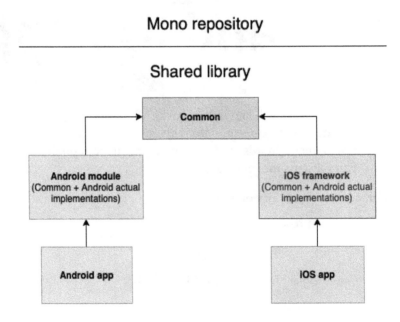

Figure 9.1 – Mono repository

Some of the KMP examples out there, such as the ones where code is shared not just between the different frontend applications but in a server-client/backend-frontend mode, could inspire people to create a more enhanced QA process of the communication layer between the client and the server:

- A communication protocol could be enforced between the backend and the frontend in the form of **data transfer objects** (**DTOs**) and endpoints defined in this shared KMP module.

- A CI pipeline would "govern" the operability of this communication protocol. If you need to make a change to the communication protocol to suit the backend, you will be able to see how this affects frontend applications.

Having something like this could shorten the feedback loop of backend changes, which usually requires direct communication between the backend and the frontend.

One other big advantage of a mono repository structure is that you can avoid the overhead of publishing the shared library for different types of consumers.

One of the biggest drawbacks is probably how it scales as having one big repository for multiple purposes may not be ideal for big projects and teams.

This option is also probably not something that published production apps would likely consider; they usually have different repositories for their different platforms and they would only share a small part of their business logic to mitigate risk.

Multiple repositories

In a multiple repository structure, the shared code is isolated from the apps and is contained in a different repository, as shown in the following diagram:

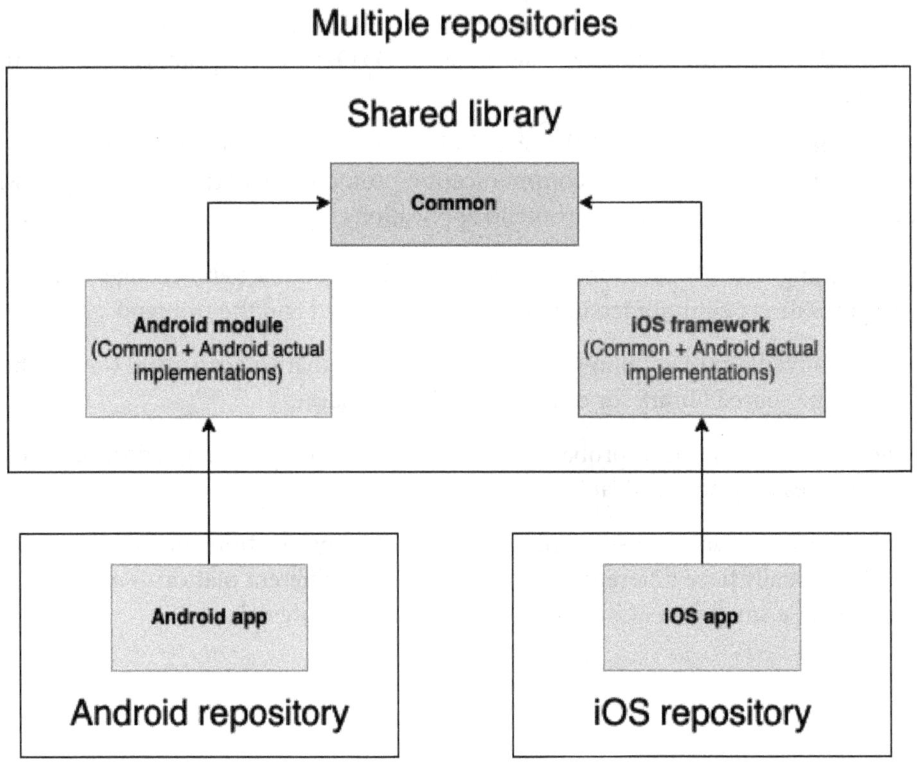

Figure 9.2 – Multiple repositories

Teams with established production apps are often more thoughtful of making big changes and refactoring, and rightly so. It makes sense for apps that are not starting on fresh grounds to experiment with KMP in a more isolated way. This helps keep the KMP world in an isolated repository that's consumed by the platforms in a platform-specific way. Here, iOS doesn't need Kotlin and/or Gradle-related tools to make sense of the shared code. They receive the library, ready to be consumed.

This setup can reduce the risk of experimenting with KMP, but at the same time has more complexity around maintaining the shared library as you need to manage an additional publishing and versioning process. For many production apps, this can be overkill.

To reduce this complexity, in the following subsections, we're going to discuss the different ways you can publish and consume a shared KMP library. We'll focus on iOS since publishing and consuming a shared KMP library is the same as for any Kotlin/JVM library, which is well documented already. In short, we can categorize the publishing formats as follows:

- **The Unreadable Shared Code**: Easy to consume but hard to read and debug

- **The Readable Shared Code**: Moderately easy to consume; easy to read and debug

- **The Modifiable Shared Code**: Harder to consume but easy to read, debug, and modify

Publishing binaries

For most teams, this approach sounds the most attractive as it isolates the KMP world and the shared library from the platform-specific Android and iOS apps.

This means that from the iOS consumer's perspective, you've chosen not to make compromises on how you consume external libraries – you'll stick with your currently preferred choice, be it Carthage, CocoaPods, or Swift Package Manager.

From the publisher's perspective, you'll need to create a binary that is well understood by your preferred dependency manager.

Fortunately, since Kotlin 1.5.30, KMP supports XCFrameworks (the replacement for fat/universal frameworks in the Apple ecosystem). We won't dive into the details of the differences between XCFrameworks and universal frameworks, but Carthage, Cocoapods, and Swift Package Manager support XCFrameworks, which means you can produce a required binary format and pull it in with your preferred dependency manager on iOS.

You can use the `assembleXCFramework` Gradle task to generate such a framework. You can find out more at `https://kotlinlang.org/docs/whatsnew1530.html#support-for-xcframeworks`.

One of the main drawbacks of publishing binaries is due to the isolation of the Kotlin world:

- It isolates the iOS team from the shared code as they view it as just another external library, while it makes the shared Kotlin code more unreachable.

- Debugging that shared code becomes problematic since the Kotlin/Native compiler adds absolute paths to files for its debugging purposes. So, binaries that have been built locally will work, but external binaries that have been pulled in won't.

This means that, in practice, you and your team are probably better off if your team has an easy way to enter the Kotlin world. This is what the guys at Touchlab worked on and the approach needs some applause.

Xcode-Kotlin plugin

You can find the Xcode-Kotlin plugin at `https://github.com/touchlab/xcode-kotlin`.

The main benefit of using this plugin is that it doesn't enforce a whole new ecosystem on iOS developers. It provides a lightweight tool on top of Xcode that they can use to debug and become familiar with the Kotlin code.

Once the plugin has been set up, it regularly clones the shared code, which can then be integrated into your current Xcode project structure. The plugin provides both debugging support and syntax highlighting for this Kotlin code.

Now, depending on your team, you may want the full experience and would like your iOS developers to become accustomed to the Kotlin ecosystem. We'll explore that option next. But in practice, it's probably best to start with the plugin first, monitor the developer experience, and if the interest is there, include them in the game even more so that they don't just "read" the shared code.

Klibs

We will be focusing on klibs in this chapter as it will probably become the standard for sharing Kotlin code in a Kotlin way.

Klib is a pure Kotlin format for distributing Kotlin code and it's a common denominator across all backends. In *Chapter 2, Exploring the Three Compilers of Kotlin Multiplatform*, we discussed the new backend compiler strategy of working with an **intermediate representation** (**IR**) of the code and producing an executable from that. This IR is the klib format. This new format standardizes how you can publish the different targets of the shared code:

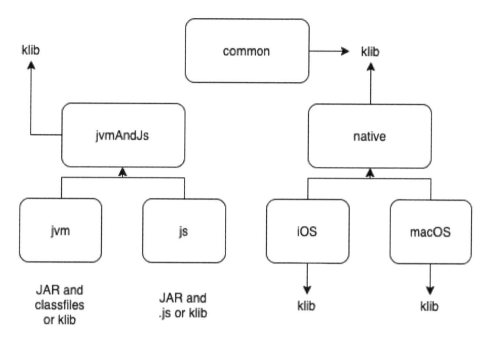

Figure 9.3 – The klib format

Kotlin bundles the target source sets with common when compiling, so jvmMain is bundled with commonMain. One of the goals of the KMP developers from JetBrains is to make source sets a proper compilation unit so that commonMain will be compiled to a separate klib that is a dependency of jvmMain. This way, you can distribute the shared code as commonMain only and plug in the supported platform as you wish.

Using this format may need the most tooling from the shared code consumers, but it will give them the most flexibility.

Conclusion

When it comes to using KMM, where the main goal is to share code between Android and iOS, you should choose a mono repository or a shared library based on the following approaches:

- **0 to 100 approach**: If you already have an Android and iOS app but would like to start sharing code in small steps, it probably makes sense not to restructure everything but to create your shared code separately and pull that in.

- **100 to 0 approach**: If you're starting fresh and would like to share as much as possible, going with a mono repository will probably be more advantageous for you.

> **Note**
> The number "100" doesn't suggest that you can get to one single code base with two apps. In reality, this will most likely top out at around 80% (maximum), but who knows what Jetpack Compose will bring – it may not only be possible but also worthwhile to write cross-platform with Jetpack Compose.

Any kind of change requires some type of adaptation, and KMP will probably have an impact on your team's structure and tooling and make some people uncomfortable. I'd argue that most of these inconveniences are similar to moving into a bigger bedroom while complaining that the light switch is far from the bed. Nevertheless, it's good to prepare yourself before you find out that, after moving into a mansion, you don't have the resources to heat it.

Exploring team structure and tooling

If you're planning on adopting KMP in your team, the following points may be obvious to you by now, but it's still worth pointing them out:

- Your shared code needs mostly Kotlin and Gradle-related expertise.
- Android teams will mostly feel natural about working with the shared code, with a relatively small amount of learning needed for KMP specifics.
- iOS teams will have a harder time, even though Kotlin and Swift are not too different. This is especially true when it comes to a new build tool, **integrated development environment** (**IDE**), and ways of working.

Team structure

Because of the aforementioned points, you should probably evaluate your team structure and plan carefully so that your shared code doesn't end up being a huge bottleneck that only a few people of your Android team will touch; it will inevitably drive your shared code toward Android and you want it to be unbiased toward platforms.

One example of a team structure I feel suits KMP is the one that the JetBrains Space team used:

- Have a dedicated team for the shared code with Kotlin and KMP expertise, focusing on the business logic. This team doesn't need expertise in the relevant platforms.

- Have dedicated platform teams (Android, iOS, and the web) that have expertise in the given platform and know the ins and outs of the specific framework.

To learn more about KMP in terms of JetBrains Space, Maxim's *Kotlin in Space* talk at the 2019 KotlinConf is a good watch: `https://www.youtube.com/watch?v=JnmHqKLgYY4&t=25m30s`.

If your team doesn't have the resources to pull off this team structure, it's fine to have the shared code experts from your Android team. Just make sure that the iOS team is part of the conversations and that they are not overly pushed to learn Kotlin and Gradle. Many good developers will probably be open to learning about them if they grasp the value proposition of KMP.

Tooling

A lot of people tend to overlook tooling, but bad or incorrectly used tooling can become a burden if it disrupts a developer's workflow. In this section, I'm going to try my best to help you to leverage the most essential tooling out there for your KMP journey.

Choosing the right IDE

If you are the Notepad/Vim + command line type, you can skip this section. Here, I'll mostly be talking about my experience with suffering through some situations in an experimental world.

So far, IntelliJ IDEA seems to be the safest choice when it comes to setting up and configuring KMP projects while offering more flexibility than just a KMM project setup. I would use IntelliJ IDEA whenever I wanted to check out a more sophisticated KMP project setup or if I was having strange build configuration issues.

Android Studio with the KMM plugin is also really promising and it is becoming better and better. For KMM applications, I'd probably use the project template offered by the KMM plugin. It can also provide a nice IDE experience for trying out and debugging the shared code on both Android and iOS. While it seems to be maturing, especially in the early days, it didn't have the stability for me to debug and run the shared code on an iOS simulator properly. Also, because syntax highlighting and code completion for Swift/Obj-C is not supported out of the box (you can improve on syntax highlighting with an IntelliJ plugin), to get the full experience on iOS, I always have Xcode set up.

So, personally, for an all-round experience of working and testing the shared code, I recommend the following IDE setup:

- IntelliJ IDEA for KMP-related build configurations. This may be needed less and less, but if you're having trouble with the setup, I'd try it out.

- Android Studio with the KMM plugin for setting up a KMM app and running/ debugging shared code on your apps.

- Xcode for the full iOS experience (Swift/Obj-C coding, managing pods, and build configurations) and possibly the Xcode-Kotlin plugin for convenient debugging on iOS.

If you want a "one" IDE experience, AppCode is a JetBrains product that may have the potential to become the KMM IDE in the future. They recently announced a KMM plugin for AppCode that lets you build, run, and debug a KMM app on both Android and iOS while having syntax highlighting and code completion capabilities for both Kotlin and Swift/Obj-C.

Debugging

We've touched on the topic of debugging your shared code a little already. On iOS, since debugging on Android isn't that different, I'd recommend either or both of the following plugins:

- The Android Studio KMM plugin or AppCode KMM plugin

- Touchlab's Xcode-Kotlin plugin, for iOS developers that want to stay as Kotlin ecosystem-free as possible

Crash reporting

Since Kotlin propagates errors/exceptions differently compared to Swift/Obj-C, shared code-related crashes may be hard to read when you're running on iOS. For this purpose, Touchlab created a nice library for making sure that these errors are still readable: `https://github.com/touchlab/CrashKiOS`.

Kevin gave a nice talk on these production questions, some of which may be or become outdated. Nevertheless, I highly recommend his talk as some of the aforementioned topics are also based on their experience: `https://www.youtube.com/watch?v=hrRqX7NYg3Q`.

In the next section, I will offer some of my personal views on how teams could leverage the benefits of KMP while lowering the risk of its adoption.

Learning some adoption tips

Have you decided that KMP is for you and your team and would like to try it out? Here are a couple of quick tips to help you mitigate risk and gradually descend into the KMP world:

- Kotlin/JVM is already used widely for Android development, which you can leverage. You can start by using Android while you introduce platform-agnostic concepts and isolate components that shouldn't depend on the Android framework.

  Many teams already do this by having Java/Kotlin modules for their business logic. These teams are already one step closer to making these modules shareable and doing this doesn't need any KMP expertise and doesn't introduce KMP specific risks.

- You can then start educating your Android team on KMP and gradually make your components platform-agnostic. Make sure that iOS is involved in the communication as they will be one of the consumers of the shared code. I'd encourage iOS people to learn about KMP and contribute, but without forcing it onto anyone.

- When you have your platform-agnostic code, you need to try out the consumer experience from iOS. This can be done iteratively, where you adapt your shared code so that consuming from iOS brings the best experience.

- If you are starting a new project, possibly for a startup, I'd push for having Android first, which has a proper architecture that can be shared with iOS. Then, you can bring that shared code to life on the iOS side.

- Since the knowledge gap between the Android world and KMP is not big, Android developers, especially ones with Kotlin knowledge, will probably find learning KMP easy. But it's not a stable framework yet, which means that some uncomfortable learning periods will occur in some scenarios and that you will need to get out of the comfort of reading well-prepared documentation.

  To shorten the learning curve and help out with undocumented issues, turning to and collaborating with already experienced KMP developers is a good shortcut.

- While KMP is not stable, you can already leverage its advantages in production apps. Many big companies already use KMP, such as Netflix, Philips, and Leroy Merlin, and you can find a list, along with case studies for each, on the official Kotlinlang website: `https://kotlinlang.org/lp/mobile/case-studies/`.

  So, the biggest question isn't "is KMP ready for production?" but "are you ready for KMP?".

And without being arrogant, what I mean by this is that the hardest thing with KMP is finding the right guidance since the road isn't that well known. In the next chapter, I'll share some of the things that worked out for me when it came to learning about KMP back in 2019, when even setting up a KMP project was pretty hard.

- Again, try not to overuse the expected/actual declarations as they are only known to the KMP world and they specify platform-specific abstractions. Using a general abstraction method such as interfaces/protocols can be much more flexible. For example, you can inject implementations from Swift and Kotlin modules that don't know about KMP-specific concepts.

Summary

In this chapter, we dived deeper into production-related questions, as well as KMP's influence on team structure.

We discussed the pros and cons of a mono repository versus a shared library and explained which one may be more suitable for you and your team.

Then, we explored how KMP could influence your team structure and the current state of the KMP ecosystem in regards to tooling. We also discussed some adoption tips that, given KMP's nature and based on your situation, may help you adopt KMP properly, with low risk and high upside.

I hope that you now have a much better picture of what to expect when you try to adopt KMP in your new or existing apps. In the next chapter, we'll go over what we've learned throughout this book and where to look for future knowledge.

10
Summary and Your Next Steps

We'll close this book by summarizing all the things that were discussed throughout and providing guidance for the next steps in your KMP learning journey.

In this chapter, I'd like to provide an overview of what we've learned in this book to emphasize the central ideas that were covered. We will then provide a quick glimpse into the future of KMP apps, which will be combined with some personal opinions, so please take them with a grain of salt. Finally, I will provide the necessary resources for you to continue your KMP journey.

This chapter will be a short one and will consist of the following topics:

- Recapitulating what you have learned
- Managing your KMP expectations
- Learning resources

Recapitulating what you have learned

We started this book by reasoning about what market need Kotlin Multiplatform tries to meet. We explored native applications, why people tend to turn to technologies such as React Native and Flutter, and why these technologies may not turn out to be the best solutions, generally speaking.

Furthermore, we discussed how to differentiate multiplatform from cross-platform notions, what a multiplatform technology tries to achieve, and how Kotlin Multiplatform leverages its tech stack to provide a sensible multiplatform solution.

After reading the first two chapters, you are expected to have a broader understanding of which technology is more suitable for a specific use case, project, or client need.

In *Chapter 3, Introducing Kotlin for Swift Developers*, we made a slight detour to ensure that iOS developers don't feel alienated by Kotlin and that everyone has the necessary knowledge to start learning about Kotlin Multiplatform.

Then, we set up our first **Kotlin Multiplatform Mobile** (**KMM**) project by writing the shared code for it, as well as the consumer Android and iOS apps, all while exploring the issues that you could come across when consuming from iOS and how to overcome them.

After that, we covered the best practices and tips in the realm of KMM and started exploring questions related to production environments, such as tooling, package management, team structure, adoption tips, and others. This has led us to this chapter, to see what we've learned and how to continue learning and exploring.

Managing your KMP expectations

This section will consist of some thoughts on the roadmap of KMP and some questions regarding its future.

Will it change the mobile development landscape?

In some sense, yes. Think of KMP as a new tool in the native app development palette that can offer good code sharing capabilities. It's already taking some "market share" from pure native and cross-platform technologies (such as React Native and Flutter). While the ratio will probably change (I'm personally bullish on KMP becoming the preferred choice), I can't see why one of these three approaches would disappear – they can all serve a specific use case.

Kotlin/Native's direct interoperability with Swift

Many people are excited about this since it's on the JetBrains teams' roadmap. I think that with the new concurrency model and direct interoperability with Swift, the usability and experience with the iOS will improve a lot, though it will probably never be perfect since Kotlin and Swift, in the end, are similar but still different in some aspects.

Tooling and documentation have improved a lot lately, probably due to the introduction of KMM and there being a dedicated KMM team at Touchlab. Still, those who are knowledgeable of infrastructure who are skilled in KMP, as well as the relevant Android and iOS build tools, will probably be needed.

Shared UI

The Compose UI toolkit is interesting and has gained a lot of traction, with more people trying to make it a toolkit that can be used on iOS as well, making it possible to write two apps with one piece of source code in KMP. This is already possible with the legacy view systems if you make the necessary abstractions on platform specifics.

For me, this is where things become cross-platform and unmaintainable as it can be hard to keep up with two frameworks.

Nevertheless, it's an interesting topic because, with KMP, you get the flexibility of sharing as much code as you want. So, you could start by sharing your business logic, then trying to share the UI-related things; if that doesn't work out, you can always go back. With other technologies, this is a different story.

Next, I'll show you where you learn more about KMP and the best ways to deepen your knowledge, at least from my experience.

Learning resources

Starting in 2019, trying to experiment with KMP was one of the best learning experiences I could've asked for. Early on, experimental stuff teaches you how best to be resourceful and to learn at a deeper level – not just by reading documentation and learning how to something, but what's behind it and how it works. It can be hard and uncomfortable at the start, but I highly recommend it for any developer.

First and foremost, I'd like to thank the huge KMP community, who helped me gain the knowledge I needed to write this book.

Although not indexed by Google, the *kotlinlang* Slack channel is a great place to start searching for any Kotlin-related issue and to start a conversation on different topics. Lately, it has been the starting point for me when searching for something, especially for experimental things. You can join the Slack channel at `https://surveys.jetbrains.com/s3/kotlin-slack-sign-up`; I highly recommend the *#multiplatform* and *#kotlin-native* channels for starters.

John O'Reilly's KMP example projects are incredible and always up to date. Whenever I was struggling with the compatibility of the different library versions, these libraries were the go-to for me, and I have the feeling that is the same for a lot of people in the community. Checking out the `https://github.com/joreilly/PeopleInSpace` repository is a good way to become familiar with his work.

Touchlab's work is also amazing in this space: `https://github.com/touchlab`. The IceRock team has also contributed with great libraries: `https://github.com/icerockdev`.

While it's impossible to list all the great contributors, you can find a list of the available KMP libraries at the following two links:

- `https://github.com/AAkira/Kotlin-Multiplatform-Libraries`
- `https://libs.kmp.icerock.dev/`

Although KMM-specific documentation was a bit lacked, it has picked up speed. JetBrains has a pretty good KMM guide at `https://kotlinlang.org/docs/kmm-overview.html`.

Last but not least, the Kotlin by *JetBrains* (`https://www.youtube.com/c/Kotlin/`) and *JetBrainsTV* (`https://www.youtube.com/user/JetBrainsTV`) YouTube channels provide a huge collection of great Kotlin talks.

Summary

In this chapter, we provided an overview of the main topics that were covered in the book, tried to reason about the future direction of KMP, and discussed some of the best ways to learn more about this technology.

I hope you've enjoyed this book and are eager to try out KMP – the community needs you! Please note that while I've tried to do my best, I have probably made some mistakes along the way; if you've noticed anything you don't agree with or require more clarity on any specific topic, don't hesitate to contact me.

Index

L

M

N

O

P

R

S

Packt.com

Subscribe to our online digital library for full access to over 7,000 books and videos, as well as industry leading tools to help you plan your personal development and advance your career. For more information, please visit our website.

Why subscribe?

- Spend less time learning and more time coding with practical eBooks and Videos from over 4,000 industry professionals

- Improve your learning with Skill Plans built especially for you

- Get a free eBook or video every month

- Fully searchable for easy access to vital information

- Copy and paste, print, and bookmark content

Did you know that Packt offers eBook versions of every book published, with PDF and ePub files available? You can upgrade to the eBook version at packt.com and as a print book customer, you are entitled to a discount on the eBook copy. Get in touch with us at customercare@packtpub.com for more details.

At www.packt.com, you can also read a collection of free technical articles, sign up for a range of free newsletters, and receive exclusive discounts and offers on Packt books and eBooks.

Other Books You May Enjoy

If you enjoyed this book, you may be interested in these other books by Packt:

How to Build Android Apps with Kotlin

Alex Forrester, Eran Boudjnah, Alexandru Dumbravan, Jomar Tigcal

ISBN: 978-1-83898-411-3

- Create maintainable and scalable apps using Kotlin
- Understand the Android development lifecycle
- Simplify app development with Google architecture components
- Use standard libraries for dependency injection and data parsing
- Apply the repository pattern to retrieve data from outside sources
- Publish your app on the Google Play store

Android UI Development with Jetpack Compose

Thomas Künneth

ISBN: 978-1-80181-216-0

- Gain a solid understanding of the core concepts of Jetpack Compose
- Develop beautiful, neat, and immersive UI elements that are user friendly, reliable, and performant
- Build a complete app using Jetpack Compose
- Add Jetpack Compose to your existing Android applications
- Test and debug apps that use Jetpack Compose
- Find out how Jetpack Compose can be used on other platforms

Packt is searching for authors like you

If you're interested in becoming an author for Packt, please visit `authors.packtpub.com` and apply today. We have worked with thousands of developers and tech professionals, just like you, to help them share their insight with the global tech community. You can make a general application, apply for a specific hot topic that we are recruiting an author for, or submit your own idea.

Share Your Thoughts

Now you've finished *Simplifying Application Development with Kotlin Multiplatform Mobile*, we'd love to hear your thoughts! Scan the QR code below to go straight to the Amazon review page for this book and share your feedback or leave a review on the site that you purchased it from.

`https://packt.link/r/1801812586`

Your review is important to us and the tech community and will help us make sure we're delivering excellent quality content.

www.ingramcontent.com/pod-product-compliance
Lightning Source LLC
Chambersburg PA
CBHW060133060326
40690CB00018B/3862